I0006437

SolidWorks Electrical 2018
Black Book

By
Gaurav Verma
&
Matt Weber
(*CADCAMCAE Works*)

Published by CADCAMCAE Works, USA. Copyright © 2018. All rights reserved. No part of this publication may be reproduced or distributed in any form or by any means, or stored in the database or retrieval system without the prior permission of CADCADCAE Works. To get the permissions, contact at cadcamcaeworks@gmail.com or info@cadcamcaeworks.com

ISBN-13 # 978-1-988722-23-8

NOTICE TO THE READER

Publisher does not warrant or guarantee any of the products described in the text or perform any independent analysis in connection with any of the product information contained in the text. Publisher does not assume, and expressly disclaims, any obligation to obtain and include information other than that provided to it by the manufacturer.

The reader is expressly warned to consider and adopt all safety precautions that might be indicated by the activities herein and to avoid all potential hazards. By following the instructions contained herein, the reader willingly assumes all risks in connection with such instructions.

The Publisher makes no representation or warranties of any kind, including but not limited to, the warranties of fitness for a particular purpose or merchantability, nor are any such representations implied with respect to the material set forth herein, and the publisher takes no responsibility with respect to such material. The publisher shall not be liable for any special, consequential, or exemplary damages resulting, in whole or part, from the reader's use of, or reliance upon, this material.

DEDICATION

To teachers, who make it possible to disseminate knowledge
to enlighten the young and curious minds
of our future generations

To students, who are the future of the world

THANKS

To my friends and colleagues

To my family for their love and support

To readers for their constructive feedbacks

Table of Contents

Chapter 3 : Line Diagram

Chapter 4 : Schematic Drawing

Preface

SolidWorks Electrical 2018 is a uniquely designed electrical CAD package from Dassault System. Easy-to-use CAD-embedded electrical schematic and panel designing enable all designers and engineers to design most complex electrical schematics and panels. You can quickly and easily employ engineering techniques to optimize performance while you design, to cut down on costly prototypes, eliminate rework and delays, and save you time and development costs. SolidWorks Electrical provides thousands of symbols and over 500,000 manufactured parts for use in your design hence saving lots of user time for designing rather than drafting.

The **SolidWorks Electrical 2018 Black Book** is the 4th edition of our series on SolidWorks Electrical software. The book is written to help professionals as well as learners in performing various tedious jobs in Electrical control designing. The book follows the best proven step by step methodology. The book covers almost all the information required by a learner to master the SolidWorks Electrical. The book starts with basics of Electrical Designing, goes through all the Electrical controls related tools and ends up with practical examples of electrical schematics. Chapters also cover Reports that make you comfortable in creating and editing electrical component reports. Some of the salient features of this book are :

In-Depth explanation of concepts

Every new topic of this book starts with the explanation of the basic concepts. In this way, the user becomes capable of relating the things with real world.

Topics Covered

Every chapter starts with a list of topics being covered in that chapter. In this way, the user can easy find the topic of his/her interest easily.

Instruction through illustration

The instructions to perform any action are provided by maximum number of illustrations so that the user can perform the actions discussed in the book easily and effectively. There are about 600 illustrations that make the learning process effective.

Tutorial point of view

The book explains the concepts through the tutorial to make the understanding of users firm and long lasting. Each chapter of the book has tutorials that are real world projects.

Project

Free projects and exercises are provided to students for practicing.

For Faculty

If you are a faculty member, then you can ask for video tutorials on any of the topic, exercise, tutorial, or concept.

Formatting Conventions Used in the Text

All the key terms like name of button, tool, drop-down etc. are kept bold.

Free Resources

Link to the resources used in this book are provided to the users via email. To get the resources, mail us at *cadcamcaeworks@gmail.com* or *info@cadcamcaeworks.com* with your contact information. With your contact record with us, you will be provided latest updates and informations regarding various technologies. The format to write us e-mail for resources is as follows:

Subject of E-mail as ***Application for resources of*** ....
book.
Also, given your information like
Name:
Course pursuing/Profession:
Contact Address:
E-mail ID:

Note: We respect your privacy and value it. If you do not want to give your personal informations then you can ask for resources without giving your information.

For Any query or suggestion

If you have any query or suggestion, please let us know by mailing us on ***cadcamcaeworks@gmail.com*** and ***info@ cadcamcaeworks.com***. Your valuable constructive suggestions will be incorporated in our books.

About Authors

The author of this book, Gaurav Verma, has authored and assisted in more than 16 titles in CAD/CAM/CAE which are already available in market. He has authored **AutoCAD Electrical Black Books** which are available in both **English** and **Russian** language. He has also authored books on various modules of Creo Parametric and SolidWorks. He has provided consultant services to many industries in US, Greece, Canada, and UK. He has assisted in preparing many Government aided skill development programs. He has been speaker for Autodesk University, Russia 2014. He has assisted in preparing AutoCAD Electrical course for Autodesk Design Academy. He has worked on Sheetmetal, Forging, Machining, and Casting in Design and Development department.

If you have any query/doubt in any CAD/CAM/CAE package, then you can contact the authors by writing at cadcamcaeworks@gmail.com

Chapter 1
Basics of
Electrical Drawings

Topics Covered

The major topics covered in this chapter are:

- *Need of Drawings*
- *Electrical Drawings*
- *Common Symbols in Electrical Drawings*
- *Wire and its Types*
- *Labeling*

NEED OF DRAWINGS

In this book, we are going through the topics related to electrical wiring and our purpose is to create good electrical wirings. So, it is important to know why we need electrical drawings and what is the role of SolidWorks Electrical in that.

When we work in an electrical industry, we need to have a lot of information handy like the wiring of machines, position of switches, load of every machine, and so on. It is almost impossible to remember all these details because there might be thousands of wires and switches, and there can be hundreds of machines in a single project. To make the information handy, we need electrical drawings that are written or printed documentation of these informations. Figure-1 shows an electrical drawing.

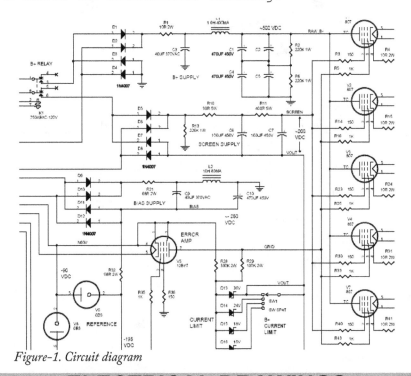

Figure-1. Circuit diagram

ELECTRICAL DRAWINGS

Electrical drawings are the representation of electrical components and connected wiring to fulfill a specific purpose. An electrical drawing can be of a house, industry or an electrical panel. An electrical drawing can be divided into following categories:

- Circuit diagram
- Wiring diagram
- Wiring schedule
- Block diagram
- Parts list

Circuit Diagram

A circuit diagram shows how the electrical components are connected together. A circuit diagram contains:

- Symbols to represent the components;
- Lines to represent the functional conductors or wires which connect the components.

A circuit drawing is derived from a block or functional diagram (see Figure-2). It does not generally bear any relationship to the physical shape, size or layout of the parts although, you can wire up an assembly from the information given in it. The circuit diagram is usually intended to show the detail of how an electrical circuit works.

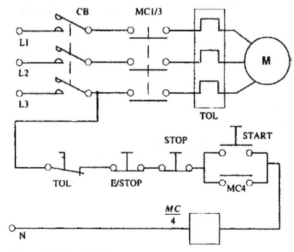

Figure-2. Circuit diagram

Wiring Diagram

This is the drawing which shows all the wiring between the parts, such as:

- Control or signal functions;
- Power supplies and earth connections;
- Termination of unused leads, contacts;
- Interconnection via terminal posts, blocks, plugs, sockets, and lead-throughs etc.

It will have details, such as the terminal identification numbers which enable us to wire the unit together. Parts of the wiring diagram may simply be shown as blocks with no indication as to the electrical components inside. These are usually sub-assemblies made separately, i.e. pre-assembled circuits or modules. Figure-3 shows a wiring diagram.

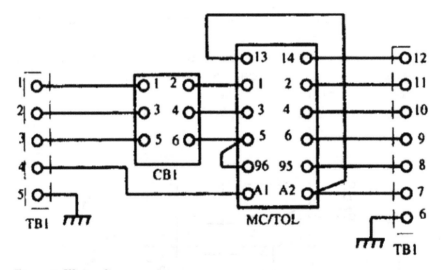

Figure-3. Wiring diagram

Wiring Schedule

A wiring schedule defines the wire reference number, type (size and number of conductors), length and the amount of insulation stripping required for soldering.

In complex equipment, you may also find a table of interconnections which will give the starting and finishing reference points of each connection as well as other important information such as wire color, identification marking and so on. Refer to Figure-4.

Schedule: Motor Control					206-A
Wire No	From	To	Type	Length	Strip Length
01	TB1/1	CB1/1	16/0.2	600 mm	12 mm
02	TB1/2	CB1/3	16/0.2	650 mm	12 mm
03	TB1/3	CB1/5	16/0.2	600 mm	12 mm
04	TB1/4	MC/A1	16/0.2	800 mm	12 mm
05	TB1/5	Ctr/1	16/0.2	500 mm	12 mm

Figure-4. Wiring Schedule

Block Diagram

The block diagram is a functional drawing which is used to show and describe the main operating principles of the equipment and is usually drawn before the circuit diagram is started.

It will not give any real detail of the actual wiring connections or even the smaller components and so is only of limited interest to us in the wiring of control panels and equipment. Figure-5 shows a block diagram.

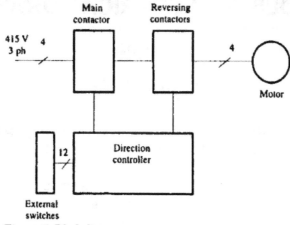

Figure-5. Block diagram

Parts list

Although not a drawing in itself, in fact it may be part of a drawing. The parts list gives vital information:

- It relates component types to circuit drawing reference numbers.
- It is used to locate and cross refer actual component code numbers to ensure you have the correct parts to commence a wiring job.

PARTS LIST			
REF	**BIN**	**DESCRIPTION**	**CODE**
CB1	A3	KM Circuit Breaker	PKZ 2/ZM-40-8
MC	A4	KM Contactor	DIL 2AM 415/50
TOL	A4	KM Overload Relay	Z 1-63

Figure-6. Parts list

You know various types of electrical drawings but these drawings contain various symbols. The following section explains the common symbols that are used in an electrical drawing.

SYMBOLS IN ELECTRICAL DRAWINGS

Symbols used in electrical drawings can be divided into various categories that are explained next.

Conductors

There are 12 types of symbols for conductors; refer to Figure-7 and Figure-8. These symbols are explained next.
1. General symbol, conductor or group of conductors.
2. Temporary connection or jumper.
3. Two conductors, single-line representation.
4. Two conductors, multi-line representation.
5. Single-line representation of n conductors.
6. Twisted conductors. (Twisted pair in this example.)

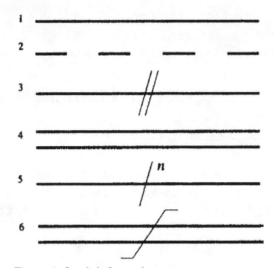

Figure-7. Symbols for conductors

7. General symbol denoting a cable.
8. Example: eight conductor (four pair) cable.
9. Crossing conductors - no connection.

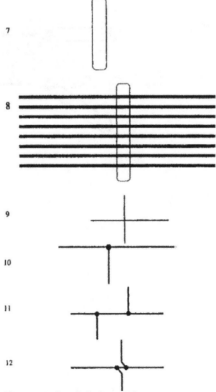

Figure-8. Symbols for conductors

10. Junction of conductors (connected).
11. Double junction of conductors.
12. Alternatively used double junction.

Connectors and terminals

Refer to Figure-9.
13. General symbol, terminal or tag.

These symbols are also used for contacts with moveable links. The open circle is used to represent easily separable contacts and a solid circle is used for those that are bolted.

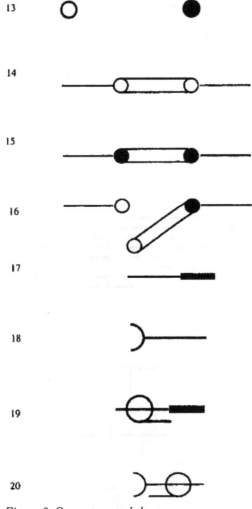

Figure-9. Connectors symbols

14. Link with two easily separable contacts.
15. Link with two bolted contacts.
16. Hinged link, normally open.
17. Plug (male contact).
18. Socket (female contact).
19. Coaxial plug.
20. Coaxial socket.

Inductors and transformers

Refer to Figure-10.
21. General symbol, coil or winding.
22. Coil with a ferromagnetic core.
23. Transformer symbols.

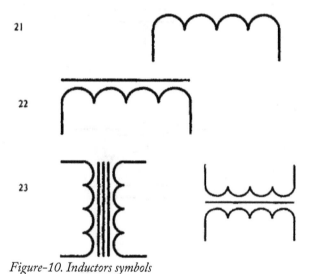

Figure-10. Inductors symbols

Resistors

Refer to Figure-11.
24. General symbol.
25. Old symbol sometimes used.
26. Fixed resistor with a fixed tapping.
27. General symbol, variable resistance (potentiometer).
28. Alternative (old).
29. Variable resistor with preset adjustment.
30. Two terminal variable resistance (rheostat).

31. Resistor with positive temperature coefficient
    (PTC thermistor).
32. Resistor with negative temperature coefficient
    (NTC thermistor).

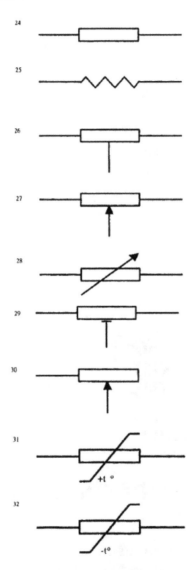

Figure-11. Resistors symbol

Capacitors

Refer to Figure-12.

33. General symbol, capacitor. (Connect either way round.)

34. Polarised capacitor. (Observe polarity when making connection.)

35. Polarized capacitor, electrolytic.

36. Variable capacitor.

37. Preset variable.

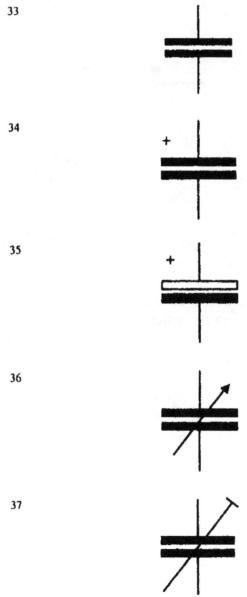

Figure-12. Capacitors symbols

Fuses

Refer to Figure-13.

38. General symbol, fuse.

39. Supply side may be indicated by thick line: observe orientation.

40. Alternative symbol (older).

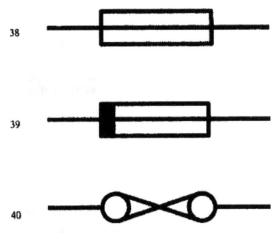

Figure-13. Fuses symbols

Switch contacts

Refer to Figure-14.

41. Break contact (BSI).

42. Alternative break contact version 1 (older).

43. Alternative break contact version 2.

44. Make contact (BSI).

45. Alternative make contact version 1.

46. Alternative make contact version 2.

47. Changeover contacts (BSI).

48. Alternative showing make-before-break.

49. Alternative showing break-before-make.

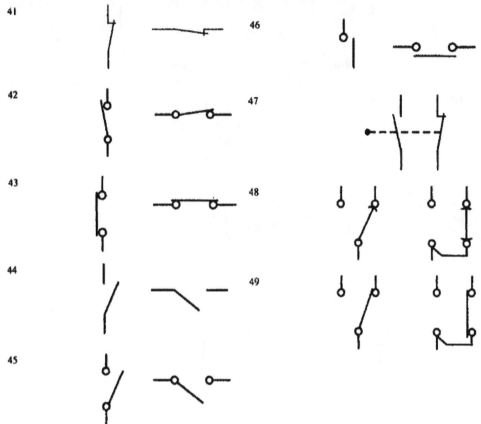

Figure-14. Switch Contact symbols

Switch types

Refer to Figure-15.
50. Push button switch momentary.
51. Push button, push on/push off (latching).
52. Lever switch, two position (on/off).
53. Key-operated switch.
54. Limit (position) switch.

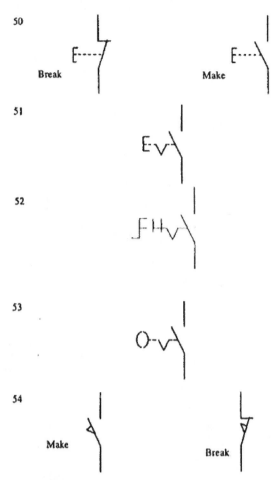

Figure-15. Switch symbols

Diodes and rectifiers

Refer to Figure-16.
55. Single diode. (Observe polarity.)
56. Single phase bridge rectifier.
57. Three-phase bridge rectifier arrangement.
58. Thyristor or silicon controlled rectifier (SCR) general symbol.
59. Thyristor - common usage.
60. Triac - a two-way thyristor.

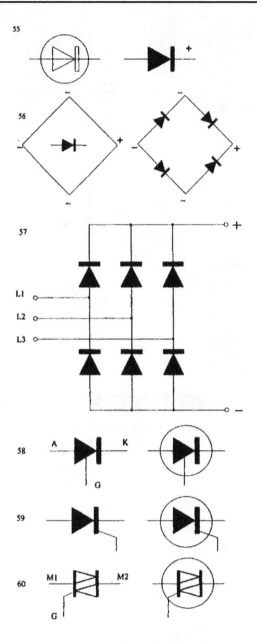

Figure-16. Diode Symbols

Earthing

Refer to Figure-17.

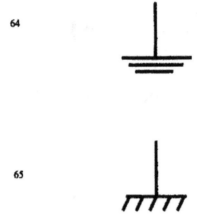

Figure-17. Earthing

Along with the above discussed symbols, you might need some user defined symbols for representation in your drawing.

After learning about various symbols the next important thing is to learn about wire and its specifications.

WIRE AND SPECIFICATIONS

Electrical equipment uses a wide variety of wire and cable types and it is up to us to be able to correctly identify and use the wires which have been specified. The wrong wire types will cause operational problems and could render the unit unsafe. Such factors include:

- The insulation material;
- The size of the conductor;
- What it's made of;
- Whether it's solid or stranded and flexible.

Types of Wires

- Solid or single-stranded wire is not very flexible and is used where rigid connections are accept able or preferred usually in high current applications in power switching contractors. It may be uninsulated.
- Stranded wire is flexible and most interconnections between components are made with it.

- Braided wire, also called Screened wire, is an ordinary insulated conductor surrounded by a conductive braiding. In this case the metal outer is not used to carry current but is normally connected to earth to provide an electrical shield to screen the internal conductors from outside electromagnetic interference.

Wire specifications

There are several ways to describe the wire type. The most used method is to specify the number of strands in the conductor, the diameter of the strands, the cross sectional area of the conductor then the insulation type.

Refer to Figure-18, Example 1:
- The 1 means that it is single conductor wire.
- The conductor is 0.6 mm in diameter and is insulated with PVC.
- The conductor has a cross-sectional area nominally of 0.28 mm .

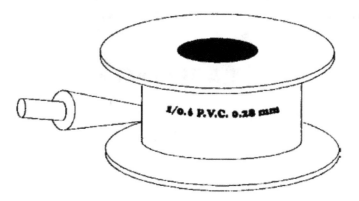

Figure-18. Example 1

Standard Wire Gauge

Solid wire can also be specified using the Standard Wire Gauge or SWG system. The SWG number is equivalent to a specific diameter of conductor; refer to Figure-19.

   For example; 30 SWG is 0.25 mm diameter.
   14 SWG is 2 mm in diameter.
   The larger the number – the smaller the size of the conductor.

There is also an American Wire Gauge (AWG) which uses the same principle, but the numbers and sizes do not correspond to those of SWG.

SWG table

SWG No.	Diameter
14 swg	2 mm
16 swg	1.63 mm
18 swg	1.22 mm
20 swg	0.91 mm
22 swg	0.75 mm
24 swg	0.56 mm
25 swg	0.5 mm
30 swg	0.25 mm

Figure-19. SWG table

We are at a position where we know about various schematic symbols and we know about wires. Now, we will learn about labeling of contactors.

LABELING

Labeling is the marking on components for identifying incoming and outgoing supply; refer to Figure-20. We also attach numbers to wires so that later on we can identify their circuits.

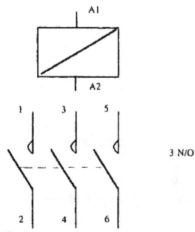

Figure-20. Contacts

Coils are marked alphanumerically, e.g. A1, A2.
   Odd numbers - incoming supply terminal.
   Even numbers - outgoing terminal.

Main contacts are marked with single numbers:
   Odd numbers - incoming supply terminal.
   Next even number - outgoing terminal.

In this way, we will find different type of markings for contacts that we would be including in our drawings.

SPACE FOR STUDENT NOTES

Chapter 2

Starting with SolidWorks Electrical

Topics Covered

The major topics covered in this chapter are:

- *Installation of SolidWorks Electrical*
- *Starting SolidWorks Electrical*
- *Project Management*
- *Archiving and Un-archiving environment*
- *Configuring Wires, Locations, and Functions*

INSTALLING SOLIDWORKS ELECTRICAL 2018

- If you are installing using the CD/DVD provided by Dassault Systemes then go to the folder containing **setup. exe** file and then right click on **setup.exe** in the folder. A shortcut menu is displayed on the screen; refer to Figure-1.

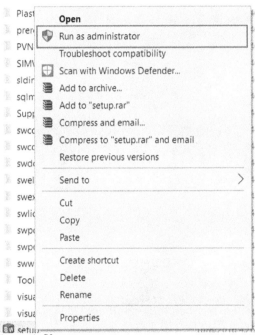

Figure-1. Shorcut menu

- Select the **Run as Administrator** option from the menu being displayed; refer to Figure 1.

- Select the **Yes** button from the dialog box displayed. The **SolidWorks Electrical 2018 Installation Manager** will be displayed. Follow the instructions given in the dialog box. Note that you must have the **Serial Number** of SolidWorks Electrical with you to install the application. Also, make sure you select the SolidWorks Electrical check box when installing. To know more about installation, double click on the **Read Me** documentation file displayed above the **setup.exe** file.

- If you have downloaded the software from Internet, then you are required to browse in the **SolidWorks Electrical Download** folder in the **Documents** folder. In this folder, open the folder of latest version available and then run **setup.exe**. Rest of the procedure is same.

STARTING SOLIDWORKS ELECTRICAL 2018

- To start SolidWorks Electrical from Start menu, click on the Start button in the Taskbar at the bottom left corner, click on the SolidWorks 2018 folder. In this folder, select the SolidWorks Electrical icon; refer to Figure-2.

Figure-2. SolidWorks Electrical option in Start menu

- While installing the software, if you have selected the check box to create a desktop icon, then you can double-click on that icon to run the software.

- If you have not selected the check box to create the desktop icon but want to create the icon on desktop, then right-click on the **SolidWorks Electrical** icon in the Start menu and select the **Send To-> Desktop (Create icon)** option from the shortcut menu displayed.

After you have performed the above steps, accept the license information. The SolidWorks Electrical 2018 application window will be display; refer to Figure-3. Since, this is the first time you are starting SolidWorks Electrical, you are asked to update the library and other data of SolidWorks. Follow the steps given next to update of libraries and SolidWorks data.

Figure-3. SolidWorks Electrical interface for first time

- Click on the **Next** button from the **Data Update** dialog box. The **Data selection** page of **Data update** dialog box will be displayed; refer to Figure-4.
- The objects in SolidWorks Electrical data that can be selected for update are displayed with their corresponding check boxes. Select the check boxes for updating corresponding objects.

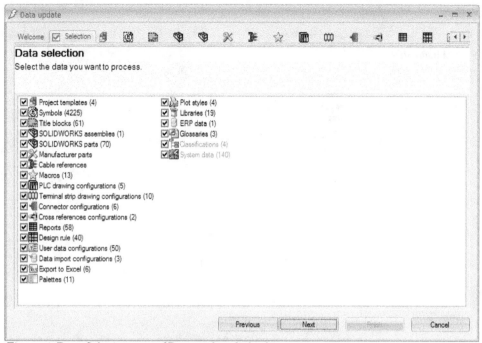

Figure-4. Data Selection page of Data update dialog box

- Note that **Add** option is selected under each category in the page. It means the Project templates will be added in the system library of current. Project template contains unit system, basic files in the form of project, title block, and other related data.
- Click on the **Next** button from the dialog box. Entities in the first selected object category will be displayed; refer to Figure-5.

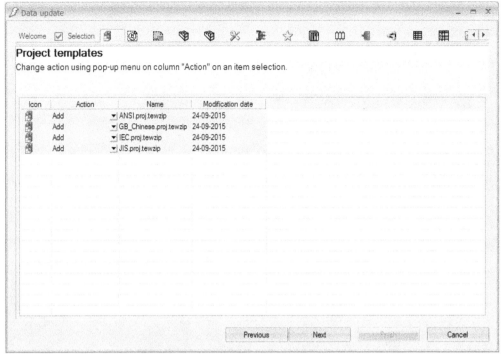

Figure-5. Objects in Project templates category

- Select the desired option for each object and click on the **Next** button. A similar page with the objects related to symbols will be displayed.
- Select the desired options and keep on clicking next button to add objects in working library. Finally, you will arrive at **Finish** page as shown in Figure-6. Summary of all the items that will be added in library is displayed in this page.
- Click on the **Finish** button from the dialog box. A report will be displayed after update is complete.
- Click on the **Finish** button to close the report. The SolidWorks Electrical interface will be displayed as shown in Figure-7.

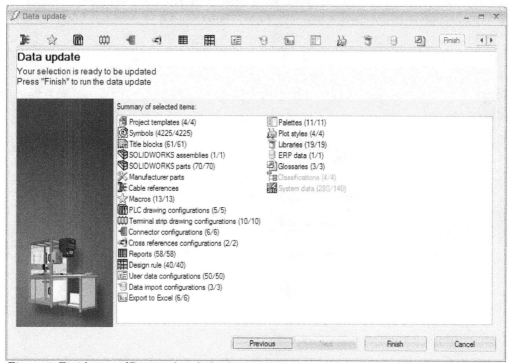

Figure-6. Finish page of Data update dialog box

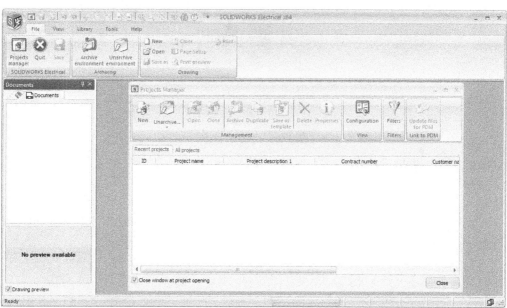

Figure-7. SolidWorks Electrical interface

Note that the first time you start SolidWorks electrical, you are asked to configure the libraries of electrical components. Like the other electrical CAD packages, SolidWorks Electrical also starts with project setup. Before starting work, you need to create a new project. The options for project setup are available in the **Project Manager** window; refer to Figure-8.

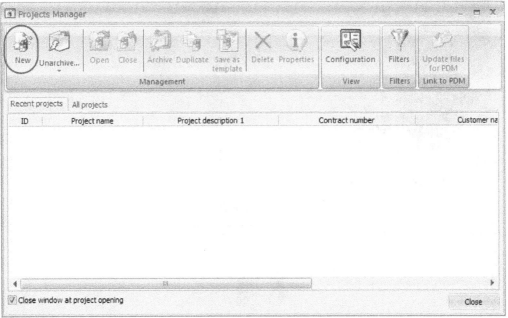

Figure-8. Project Manager

The details of **Project Manager** window are discussed next.

PROJECT MANAGER

Project Manager is used to perform all the general operations related to projects like, stating a new project, opening an existing project, creating copies of project files, and so on. Various operations that can be done in **Project Manager** are discussed next.

Starting a new project

This is the most crucial step for creating electrical drawings in SolidWorks Electrical as all the successive parameters depend on this step. At this step, we decide the electrical standards to be used during the creation of electrical drawings. The steps to start a new project are given next.

- Click on the **New** button from the **Ribbon** in the **Project Manager** window; refer to Figure-8. The **Create a new project** dialog box will be displayed with various electrical standards that can be used; refer to Figure-9.

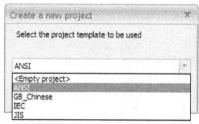

Figure-9. Create a new project dialog box

- Select the desired option from the drop-down in the dialog box and click on the **OK** button. (We have selected ANSI in our case.)
- On doing so, the files related to selected standard will be loaded. If you have selected multiple languages during setup, then **Project language** dialog box will be displayed; refer to Figure-10.

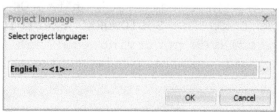

Figure-10. Project language dialog box

- Select the desired language and click on the **OK** button (We have selected **English** in our case). On doing so, the **Project** dialog box will be displayed; refer to Figure-11.
- Specify the desired name in the **Name** edit box; refer to Figure-11

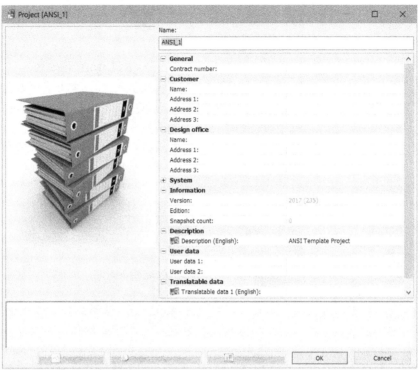

Figure-11. Project dialog box

- One by one, click in the fields of table and specify the data related to customers.
- After specifying the data, click on the **OK** button from the **Project** dialog box. The **SOLIDWORKS Electrical** information box will be displayed notifying you that the database is getting connected for the project; refer to Figure-12.

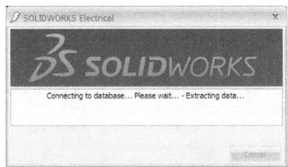

Figure-12. SOLIDWORKS Electrical information box

- Once the process of database linking is completed, the auto-generated document set will be displayed in the **Document Browser** available in the left of application window; refer to Figure-13.

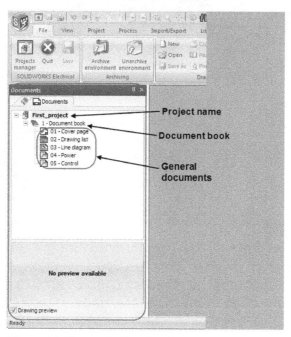

Figure-13. Documents Browser

We will discuss more about **Document Browser** later in the chapter. Now, we will discuss other options in the **Project Manager** window.

Un-archiving projects

SolidWorks Electrical provides options to archive projects that are not in use currently. These projects, after archiving, are stored at the server or in local memory. The **Unarchive** option in the **Project Manager** allows us to unpack those stored projects. The steps to un-archive a project are given next.

- Click on the **Unarchive** tool from the **Project Manager** window. The **Open** dialog box will be displayed and you will be asked to select a SolidWorks Electrical archive file; refer to Figure-14.

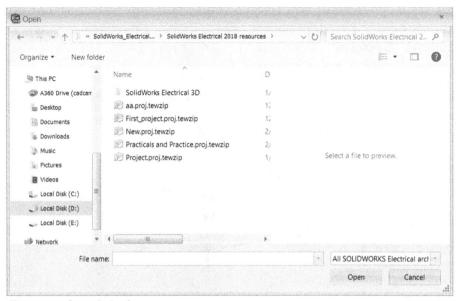

Figure-14. Open dialog box

- Select the archive file that you have earlier saved in your record and click on the **Open** button from the dialog box. The **SolidWorks** information box will be displayed notifying you that the file is being extracted. Once the extraction is complete, the **Project** dialog box will be displayed showing the basic information of the project extracted; refer to Figure-15.

- Click on the **OK** button from the **Project** dialog box to add the extracted project in the **Project Manager** window. The **Merge library elements** dialog box will be displayed asking you whether to update the library with symbols of extracted project or not; refer to Figure-16.

Figure-15. Project dialog box with information of extracted project

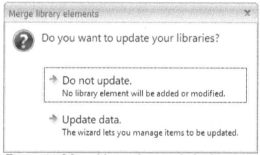

Figure-16. Merge library elements dialog box

- Click on the **Do not update** option from the dialog box and you are asked whether to open the documents of project or not. Open the project to work on it. If you have selected the **Update data** option from the dialog box then **Unarchiving: Projects** dialog box will be displayed; refer to Figure-17.

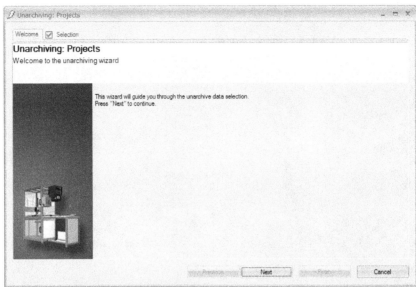

Figure-17. Unarchiving: Projects dialog box

- Click on the **Next** button from the dialog box and follow the instructions given in dialog box.
- When you have included all the desired components, click on the **Finish** button from the dialog box; refer to Figure-18.

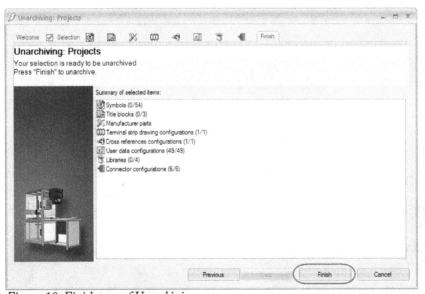

Figure-18. Finish page of Unarchiving

- Now, open the project to work on it.

Archiving a project

Archiving of a project is done to save the data related the project in a compresses folder. This compressed folder can be shared with the peers and customers for further modifications. In previous topic, we have un-archived a project and now we will do the reverse.

- If the project which you want to archive is open then select the project from the **Project Manager** and click on the **Close** button to close it; refer to Figure-19. On doing so, the project details will be displayed in black color which were bold blue earlier.

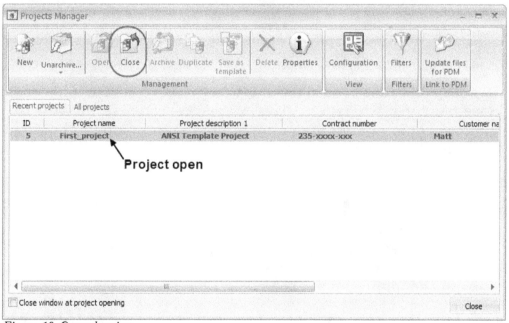

Figure-19. Opened project

- Click on the **Archive** tool from the **Ribbon** in the **Project Manager**. The **Save As** dialog box will be displayed, prompting you to save the archive file; refer to Figure-20.

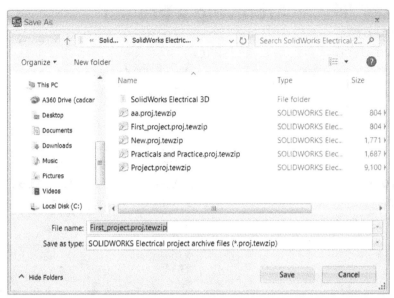

Figure-20. Save As dialog box

- Specify the desired name of the archive and click on the **Save** button to save the file. Once the process of archiving is complete, you will be asked whether to open the folder (in which archive is saved) or not.
- Choose **Yes** or **No** are required.

Duplicating Project

Using the **Duplicate** button in the **Project Manager**, you can create duplicate of a project selected in **Project Manager**. On choosing this button, you will be asked to specify new name for the duplicate project.

Saving as template

Any of the project you have created earlier can be used as template for successive projects. To make a project as template, follow the steps given next.

- Click on the **Save as Template** button from the **Project Manager** after selecting a project. The **Project** dialog box will be displayed as shown in Figure-21.

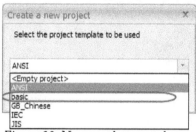
Figure-21. Project dialog box

- Specify the name of the template in the **Name** edit box and click on the **OK** button from the dialog box. Once the processing is complete, you can see the new template in the template list used for creating new projects; refer to Figure-22.

Figure-22. New template created

If you have a long list of projects in the database, use the **Filters** button to filter out the one on which you want to work.

If you work with the peers using the PDM workgroup then click on the **Update files for PDM** button to update file on PDM workgroup so that others can see the latest changes.

Project Configuration

We are not talking about the **Configuration** tool in the **View** panel in the **Project Manager**. Here, we are talking about configuring a project. The tool to configure a project is available in the **Project** dialog box displayed after selecting **Properties** button from the **Project Manager**; refer to Figure-23.

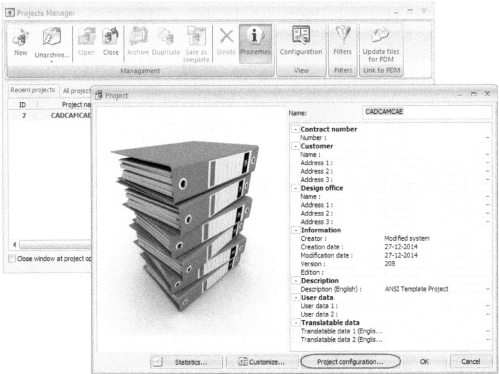

Figure-23. Project configuration button

- Click on the **Project configuration** button from the **Project** dialog box. The **Project configuration** dialog box will be displayed as shown in Figure-24.

General tab options

- Using the options in the **Project languages** node, you can specify the main language and alternate languages for the project.
- Select the desired standard for unit and cable sizes by using the options in the **Standard** node.
- Using the options in the **Date display format** node, you can change the format of date displayed in the drawings of current project.
- Similarly, you can change the revision numbering format by using the **Format** field in the **Revision numbering** node.
- Click in the **Book** field under the **<Default>** node to change the default book of project. Once you have changed the default book, all the new drawings will be automatically added in the selected default book.

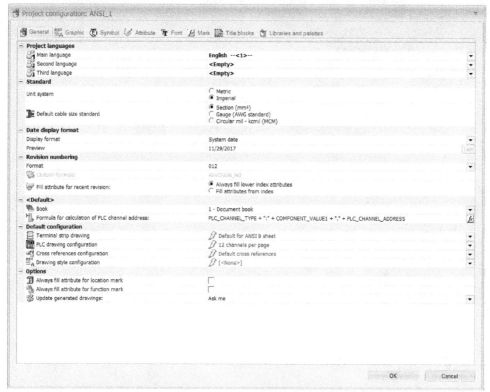

Figure-24. Project configuration dialog box

- Click on the **fx** button in the **Formula for calculation of PLC channel address** field to change the PLC address calculation formula; refer to Figure-25. The **Formula manager** will be displayed as shown in Figure-26.

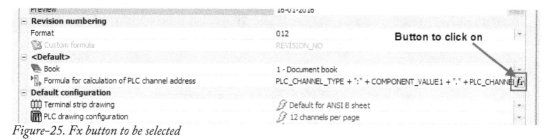

Figure-25. Fx button to be selected

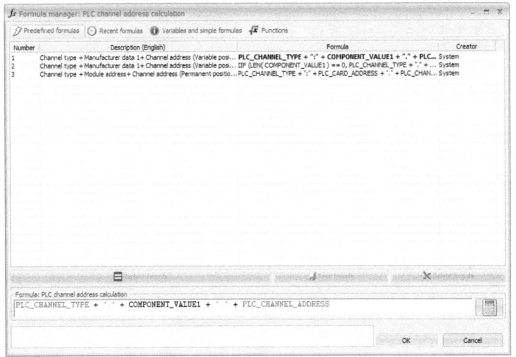

Figure-26. Formula manager

- Select the desired formula from the list or add a new formula and select it. Click on the **OK** button from the dialog box to apply formula.
- From SolidWorks Electrical 2017 onwards, you can make your drawings update automatically by using the **Always** option from the **Update generated drawings** drop-down in the dialog box; refer to Figure-27. Note that if you now close the project file or SolidWorks Electrical then all the generated drawings will be updated automatically.

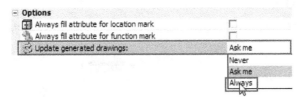

Figure-27. Update generated drawings options

- Similarly, you can set the other options of project.

Graphic tab options

- Click on the **Graphic** tab from the dialog box to display the options related to graphics; refer to Figure-28.
- Click on the color, value, or line type to change it for the desired entity.

Symbol tab options

- You can modify the label of cable, wire, location and equipotential by using the **Select**, **Remove**, or **Modify** buttons displayed at the bottom of each symbol in the dialog box; refer to Figure-29.

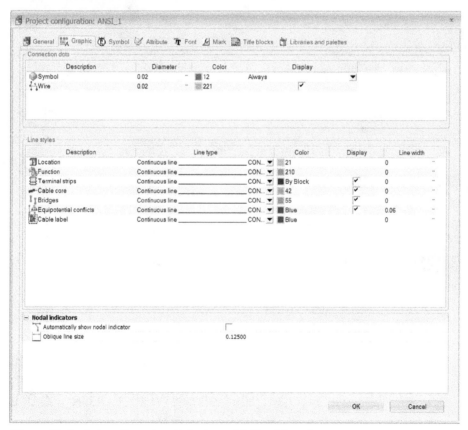

Figure-28. Project configuration dialog box with Graphic options

- If you wish to modify a tag then click on the modify button for that label in the dialog box. The respective label will open in SolidWorks Electrical in the background.
- Exit the dialog boxes by press ESC from keyboard. The selected label will be displayed with options to edit; refer to Figure-30.

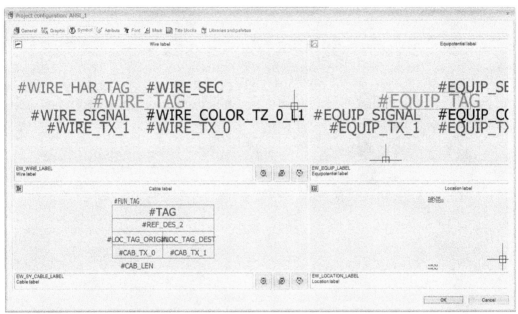

Figure-29. Buttons to modify labels

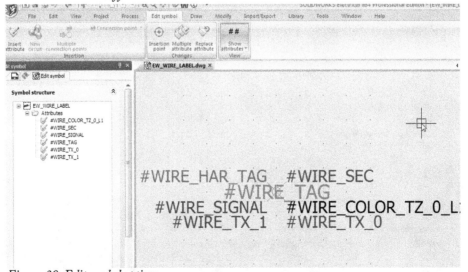

Figure-30. Edit symbol options

Attribute tab options

- Click on the **Attribute** tab from the dialog box to display the options related to attributes; refer to Figure-31.

Figure-31. Project configuration dialog box with Attribute options

- Click on the **Add custom attribute** button to add new custom attribute for all the other parameters. Set the desired parameters.
- Click on the **Add symbol attribute** button to add new attribute for new symbols. The **Attribute Management** dialog box will be displayed; refer to Figure-32. Select the desired parameters and click on the **OK** button.

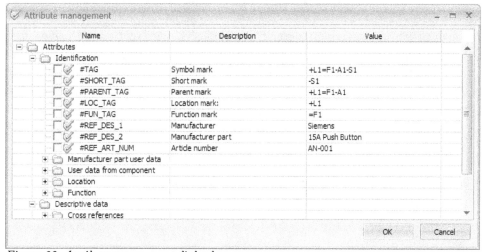

Figure-32. Attribute management dialog box

- Similarly, you can set the other parameters in the **Attribute** tab.

Font tab options

- Click on the **Font** tab from the dialog box to display the options related to fonts; refer to Figure-33.

Figure-33. Project configuration dialog box with Font options

- Click on the field under **Font** column and select the desired language for the respective object.
- Click on the field under the **Height** column and change the height as per the requirement. Similarly, you can change the other values related to font at the bottom in the dialog box. Note that you can also define formulae for **Scheme cable core** and **Line diagram cable** using the **Fx** option next to their formula field. You will learn more about formulae later in the book.

Mark tab options

- Click on the **Mark** tab from the dialog box to display the options related to various markings; refer to Figure-34.

Figure-34. Project configuration dialog box with Mark options

- Using the options in this page, you can change the default marking method for various entities like location marking, cable marking, and so on.

Similarly, you can modify, title block and libraries by using the respective page in the dialog box. Note that if you want to change the symbol palette style from **ANSI** to **IEC** then you can change it using the options in **Libraries and palettes** tab of the dialog box.

- After setting the desired parameters, click on the **OK** button from the dialog box.

ARCHIVING ENVIRONMENT

As the name suggests, archiving environment means archiving all the library of symbols, components and standards used in the project. This archive can be shared with the peers who do not have the database updated by you. For example, you have created around 100 new symbols that are specific to your organization and you want to work with another person who is at different location. Then, you can archive the environment and send it to him so that he has the updated symbol library. The procedure to archive environment is given next.

- Click on the **Archive environment** tool from the **Archiving** panel in the **Ribbon**. The **Archiving: Environment** dialog box will be displayed; refer to Figure-35.

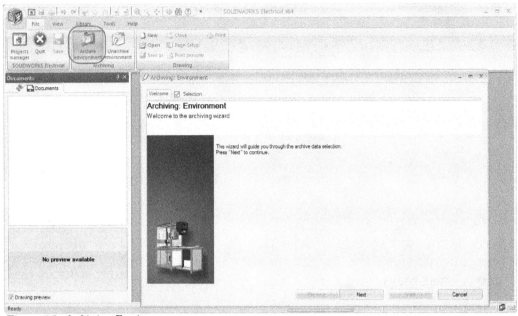

Figure-35. Archiving Environment

- Click on the **Next** button from the dialog box to make selections of the entities. The dialog box will be displayed as shown in Figure-36.
- By default, all the objects are selected for archiving. Click on the **Custom** radio button to select the objects as per your requirement.
- After selecting the desired objects, click on the Next button from the dialog box. A summary page will be displayed.
- Click on the **Finish** button from the dialog box. The **Save As** dialog box will be displayed asking you to save the archive file.
- Specify the desired name of the archive and save it at the desired location. A dialog box with report will be displayed. Click on the **Finish** button to exit the dialog box displayed.

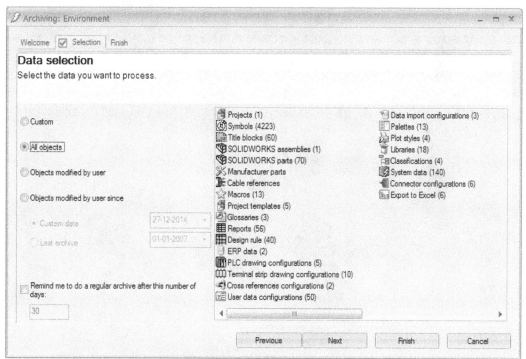

Figure-36. Data Selection page of Archiving Environment

UNARCHIVE ENVIRONMENT

The **Unarchive environment** tool is available in the **Archiving** panel of the **Ribbon**. The **Unarchive environment** tool makes reversal of **Archive environment** tool. It works in the same way as **Unarchive** tool in the **Project Manager**.

ADDING NEW BOOK/FOLDER IN PROJECT

When we are working on projects that have hundreds of drawings then we generally categorize the files on the basis of their functioning. For example, we are working on the a project which has the drawings of electrical distribution of a city. There are n number of companies, houses, and commercial parks; each having their own electrical distribution drawing. So, the drawings of various houses, commercial parks, and companies in the same locality are placed under a book/ folder having name of the locality.

Adding New Book

The procedure to add a new book in the project is given next.

• Click on the **New book** tool from the **New** drop-down in the **Project** panel of the **Project** tab in the **Ribbon**; refer to Figure-37. The **Book** dialog box will be displayed; refer to Figure-38.

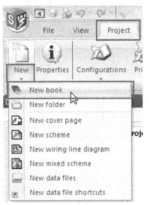

Figure-37. New book tool

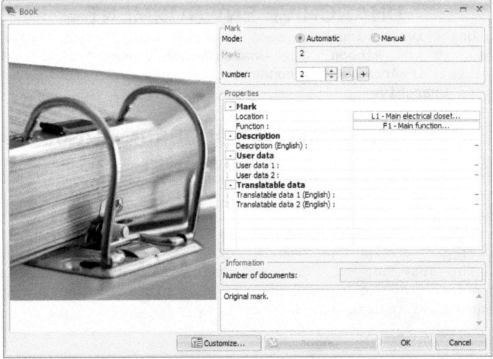

Figure-38. Book dialog box

- A mark number is automatically added to the book which is **2** in our case; refer to **Mark** area in the dialog box shown in Figure-38.
- If you want to manually specify the marking then click on the **Manual** radio button at the top in the dialog box and specify the desired mark parameter in the **Mark** edit box; refer to Figure-39. In this figure, we have specified the marking as Street 5- Block 1 which is abbreviated is **St.5-1**.

Figure-39. Manual Marking

- Click in the field adjacent to **Location** in the **Properties** box of the dialog box. The **Select location** dialog box will be displayed; refer to Figure-40.

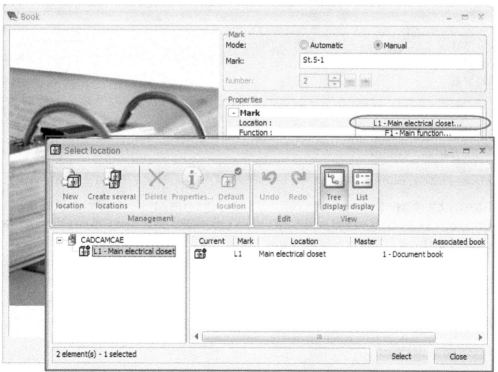

Figure-40. Select location dialog box

- Location is the position of component in the electrical closet. We will learn more about the location later in the chapter.
- Select the desired location and click on the **Select** button from the dialog box displayed.
- Similarly, select the desired function from the Select function dialog box displayed on clicking in the field adjacent to **Function** in the **Properties** box of the dialog box; refer to Figure-41.

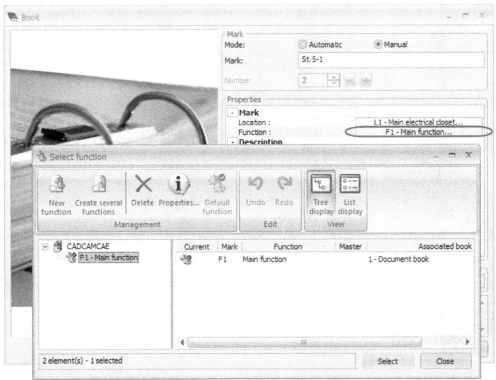

Figure-41. Select function dialog box

- Specify the description and other user information in the **Properties** box and click on the **OK** button from the **Book** dialog box to create the book.

Adding New Folder

The procedure to add a new folder in the project is given next.

*   Click on the **New folder** tool from the **New** drop-down in the **Project** panel of the **Project** tab in the **Ribbon**. The **Folder** dialog box will be displayed; refer to Figure-42.

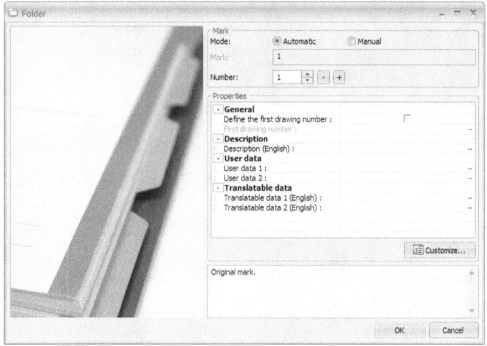

Figure-42. Folder dialog box

*   Most of the options in this dialog box are same as discussed for the **Book** dialog box.

- In the **Folder** dialog box, select the **Define the first drawing number** check box and specify the desired number for the first drawing in the folder if you want to.
- Click on the **OK** button to create the folder. The folder will be added in the selected book; refer to Figure-43.

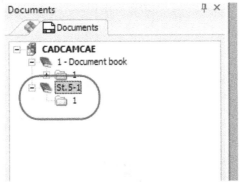

Figure-43. Folder added in the book

ADDING NEW DRAWING IN PROJECT

Cover page, Line diagram, Schema and Mixed Scheme are collectively called drawings in SolidWorks Electrical. The tools to add these drawings are also available in the **New** drop-down in the **Project** panel of the **Project** tab in the **Ribbon**; refer to Figure-44. The procedure to start any of these drawings is similar. Here, we will discuss the procedure to add a new schematic drawing (schema). You can add the other drawings on the basis of that.

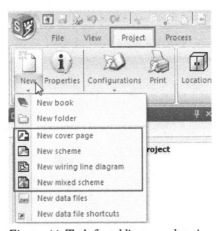

Figure-44. Tools for adding new drawing

Adding Schema

The procedure to add a schema in the project is given next.

- Click on the **New schema** tool from the **New** drop-down in the **Project** panel of the **Project** tab in the **Ribbon**. The **Drawing** dialog box will be displayed as shown in Figure-45.

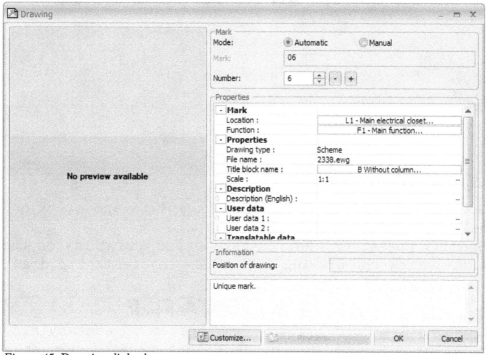

Figure-45. Drawing dialog box

- Mark number is automatically assigned to the drawing since the **Automatic** radio button is selected in the **Mark** area of the dialog box. In our case, the mark number is **6**. To give a user-defined number, select the **Manual** radio button and specify the desired mark number.
- Set the desired scale for drawing by clicking in the field corresponding to **Scale** in the dialog box.
- Enter the description about the drawing in the **Description (English)** field.
- Click in the field corresponding to **Title block name**. The **Title block selector** dialog box will be displayed as shown in Figure-46.

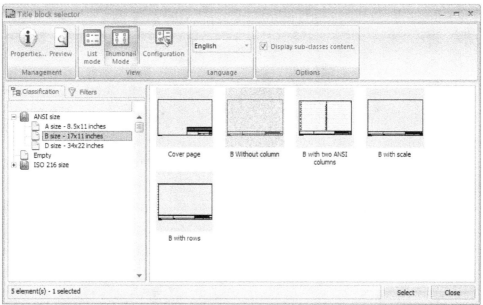

Figure–46. Title block selector dialog box

- Select the desired title block template and click on the **Select** button from the dialog box.
- Specify the other desired parameters and click on the **OK** button from the **Drawing** dialog box to create the drawing.

ADDING A DATA FILE

SolidWorks Electrical gives you freedom to add any file you want as data file in the project. Note that you must have a program installed in your system to open the data file because SolidWorks uses the default Windows program to open the data file. I know some of the people will add mp3 file as data file!! Note that once you add a file as data file to SolidWorks Electrical Project then a copy of the file is stored with the project files. So if you change the data file in SolidWorks Electrical then it will not be reflected in the original data file. Procedure to add a data file is given next.

- Click on the **Add data files** tool from the **New** drop-down in the **Project** panel of the **Project** tab in the **Ribbon**. The **Open** dialog box will be displayed as shown in Figure-47.

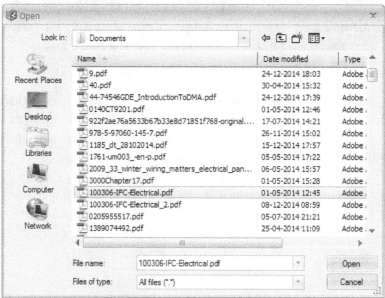

Figure-47. Open dialog box

- Select the desired file and click on the **Open** button from the dialog box. The selected file will be added in the current book of the project.

ADDING DATA FILE SHORTCUT

Okay! we have added a data file in SolidWorks project then why do we need a data file shortcut. The answer is: When you add a data file shortcut then on changing the file in SolidWorks Electrical, the original file also changes. This makes the process more streamlined. The procedure to add data file shortcut is given next.

- Click on the **New data file shortcuts** tool from the **New** drop-down in the **Project** panel of the **Project** tab in the **Ribbon**. The **Open** dialog box will be displayed as shown in Figure-47.
- Double-click on the file whose shortcut is to be added in the project. A shortcut file will be added in the project.

CONFIGURING WIRES

Wires are the life-line of any circuit. It is important to configure wires before using them. The procedure to configure wires is given next.

• Click on the **Wire styles** option from the **Configurations** drop-down in the **Project** panel of the **Project** tab in **Ribbon**; refer to Figure-48. The **Wire style manager** will be displayed; refer to Figure-49.

Figure-48. Wire styles option

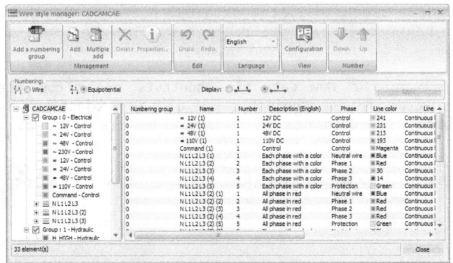

Figure-49. Wire style manager

- To modify the wire style, double-click on it in the table. The **Wire style properties** dialog box will be displayed; refer to Figure-50.

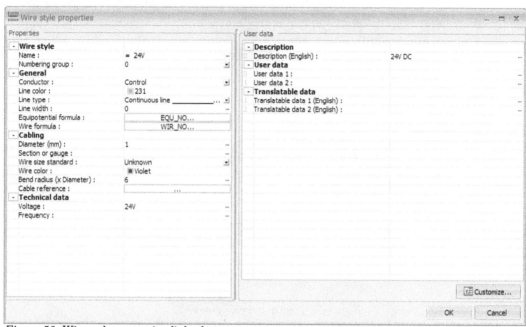

Figure-50. Wire styles properties dialog box

- Click in the desired field and change the parameters as per your requirement.
- Click on the **OK** button to set the wire style properties.

Adding a Numbering Group

A numbering group is used to categorize wires as per their usage. In SolidWorks Electrical, wires with equipotential counter are placed in one group. The procedure to create a numbering group is given next.

- Click on the **Add a numbering group** tool from the **Wire style manager**; refer to Figure-51. The **New numbering group** dialog box will be displayed as shown in Figure-52.

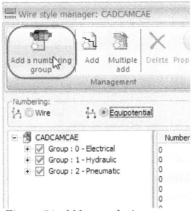

Figure-51. Add a numbering group button

Figure-52. New numbering group dialog box

- Specify the desired number in the edit box available in the dialog box and click on the **OK** button. A new group will be added in the **Wire style manager**; refer to Figure-53.

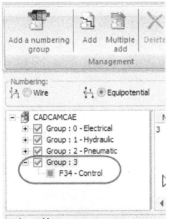

Figure-53. New wiring group

- Right-click on the name of newly created group. A shortcut menu will be displayed.
- Select the **Properties** option from the shortcut menu; refer to Figure-54. The **Numbering group** dialog box will be displayed; refer to Figure-55.

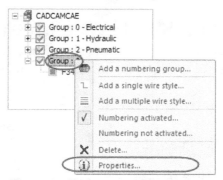

Figure-54. Shortcut menu for wire group

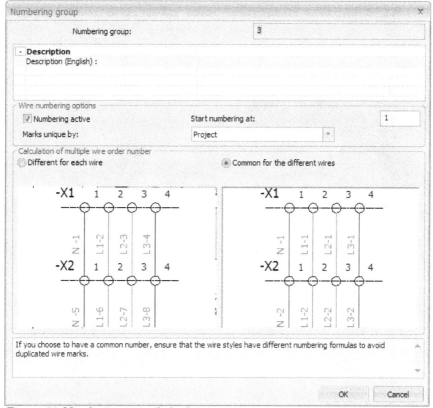

Figure-55. Numbering group dialog box

- Click in the field adjacent to **Description (English)** in the table and specify the desired description for the group.
- Specify the numbering and marking scheme for the wires by using the other options in the dialog box and then click on the **OK** button from the dialog box.

Adding Single Wire in Group

- Click on the **Add** button from the **Management** panel in the **Wire style manager.** A new wire will be added in the selected group.
- Double-click on the wire in the table. **Wire style properties** dialog box will be displayed as discussed earlier. Change the properties as required.

Adding Multiple Wire in Group

- Click on the **Multiple Add** button from the **Management** panel in the **Wire style manager.** A set of multiple wires will be added in the group; refer to Figure-56.

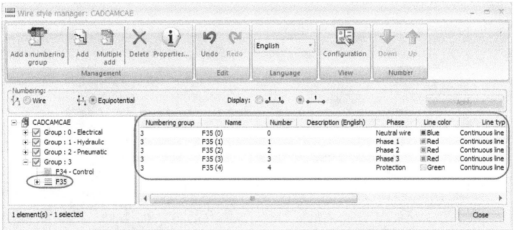

Figure-56. Set of multiple wires

- Expand the node to check the individual wires. Double-click on each wire to change its properties.

- Use the **Up** and **Down** buttons in **Number** panel of the **Wire style manager** to change the position and numbering of wire in the **Wire style manager**.
- To delete any wire style, select it and press **DELETE** button from the keyboard.

We will discuss about the other configurations in their related chapters, later in the book. Now, we will understand the concept of location and function.

LOCATIONS

Locations are used to group the components on the basis of their locations in circuit, panel, or placement in floor plan. For example, you have a common panel for three storey building. Then you can define the locations as L0 for base floor, L1 for first floor and so on. Locations help to identify the components. The procedure to create location codes is given next.

- Click on the **Locations** tool from the **Management** panel in the **Project** tab of the **Ribbon**. The **Locations manager** will be displayed; refer to Figure-57.

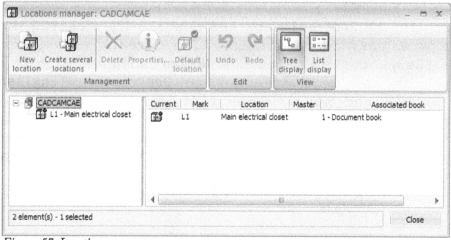

Figure-57. Locations manager

- **L1** is available in the **Locations manager** by default. To add more locations, click on the **New location** tool from the **Management** panel in the **Locations manager**. The **Location** dialog box will be displayed as shown in Figure-58.

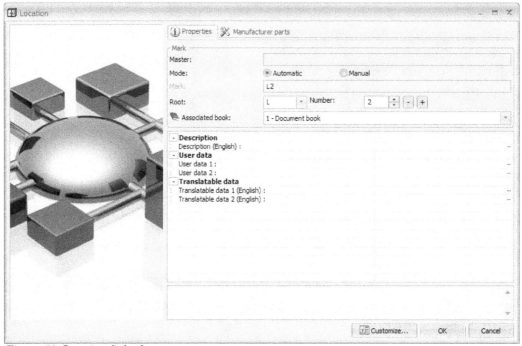

Figure-58. Location dialog box

- Click in the **Root** edit box and specify the desired keyword for your easy identification like, **Area**.
- Set the number by using the **Number** spinner/edit box.
- Click in the field adjacent to **Description (English)** in the table and specify description about the location. Similarly, set the other user data; refer to Figure-59.
- After setting the data, click on the **OK** button from the dialog box. The new location will be added in the **Locations manager**.

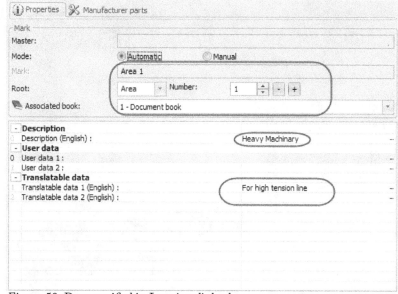

Figure-59. Data specified in Location dialog box

Creating Multiple Sub-locations

- Click on the **Create several locations** tool from **Management** panel in the **Locations manager**. The **Multiple insertion** dialog box will be displayed; refer to Figure-60.

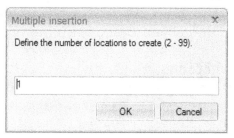

Figure-60. Multiple insertion dialog box

- Specify the desired number of sub-locations that you want to add in the selected location and click on the **OK** button. The sub-locations will be added under the location; refer to Figure-61.

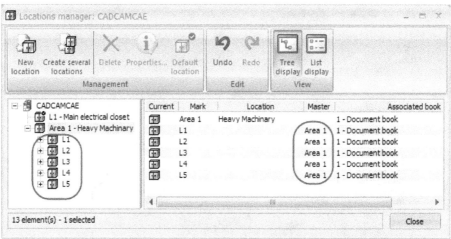

Figure-61. Sub-locations added

- Double-click on the location in table to change its properties.

FUNCTIONS

Functions are used to identify the components on the basic of their collective function. For example, there are 10 components that are used to control the motors then these components can be put under the function named control. The procedure to add functions is given next.

- Click on the **Functions** tool from the **Management** panel in the **Project** tab of the **Ribbon**. The **Functions manager** will be displayed as shown in Figure-62.

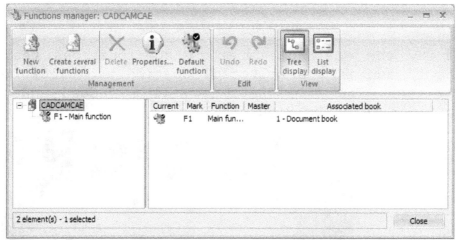

Figure-62. Functions manager

- Click on the **New function** button from the **Management** panel in the **Functions manager**. The **Function** dialog box will be displayed; refer to Figure-63.

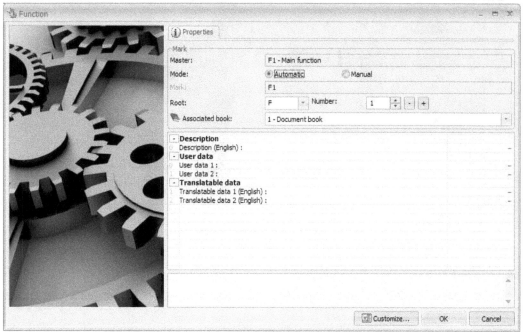

Figure-63. Function dialog box

- Specify the parameters as done for location in the **Location** dialog box.
- Click on the **OK** button from the dialog box. The function will be added in the **Functions manager**.

Creating Multiple Sub-functions

- Click on the **Create several functions** button from the **Management** panel in the **Functions manager**. The **Multiple insertion** dialog box will be displayed; refer to Figure-64.

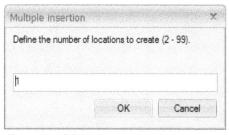

Figure-64. Multiple insertion dialog box

- Specify the number of sub-functions that you want to add in the edit box and click on the **OK** button from the dialog box. The sub-functions will be added in the selected function; refer to Figure-65.

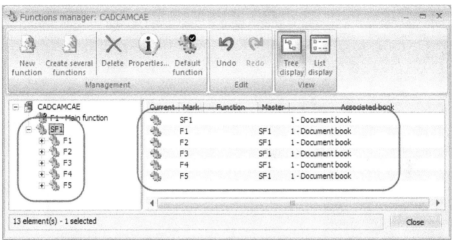

Figure-65. Sub-functions added

- Double-click on the functions to change their properties.

Chapter 3

Line Diagram

Topics Covered

The major topics covered in this chapter are:

- *Creating Line diagrams*
- *Inserting Symbols and Manufacturer parts*
- *Connection labels*
- *Drawing cables*
- *Origin - Destination Arrows*
- *Function outline and Location outline*
- *Detailed Cabling*

INTRODUCTION

Line Diagrams are used to represent the complete cabling with the help of single lines and components. In the case of line diagrams, we don't have to insert detailed schematics. We insert only simplified representations of component to displayed the cabling arrangement.

CREATING LINE DIAGRAM

*   Click on the **New wiring line diagram** tool from the **New** drop-down in the **Project** panel of the **Project** tab in the **Ribbon**. The **Drawing** dialog box will be displayed with the options related to line diagram; refer to Figure-1. If the dialog box is not displayed by default then right-click on the newly added drawing from the **Documents browser** and select the **Properties** button; refer to Figure-2.

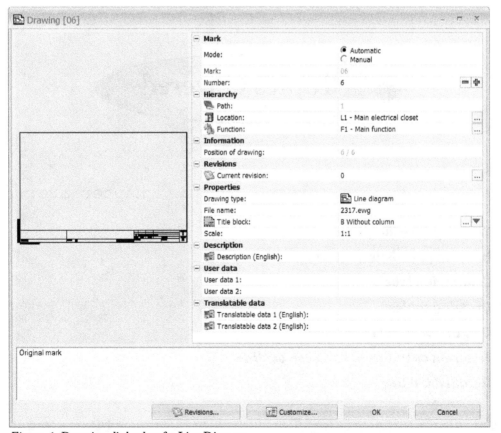

Figure-1. Drawing dialog box for Line Diagram

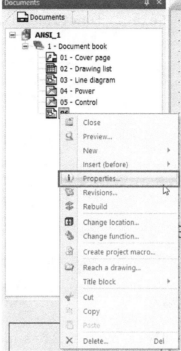

Figure-2. Shortcut menu for properties

- Specify the drawing number in the **Number** edit box or select **Manual** radio button and specify the desired identifier.
- Click in the **Location** field and set the location of drawing from the **Select location** displayed; refer to Figure-3.

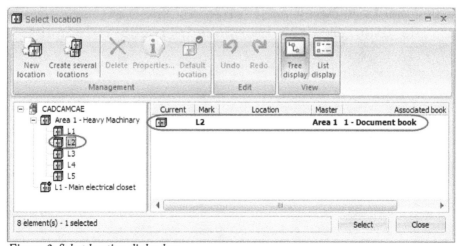

Figure-3. Select location dialog box

- Click in the **Function** field and set the function of the drawing from the **Select function** dialog box; refer to Figure-4.

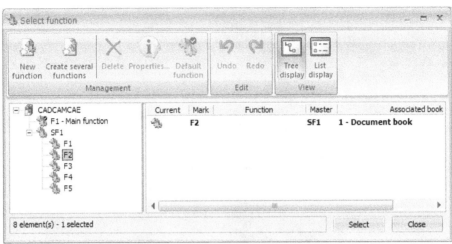

Figure-4. Select function dialog box

- Click in the **Title block name** field and set the title block for the drawing from the **Title block selector** dialog box.
- Click in the **Description (English)** field and specify the description about the drawing.
- Click on the **OK** button from the **Drawing** dialog box. The drawing will be added in the project.
- Double-click on the newly added line diagram drawing from the **Documents browser** available at the left of the window. The line diagram will be opened; refer to Figure-5.

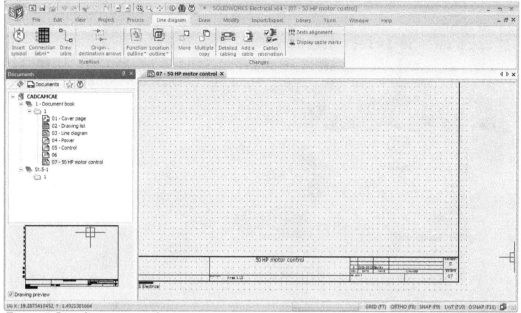

Figure-5. Line diagram opened

INSERTING SYMBOLS

- Click on the **Insert symbol** button from the **Insertion** panel in the **Line diagram** tab in the **Ribbon**. The **Symbol selector** dialog box will be displayed as shown in Figure-6.

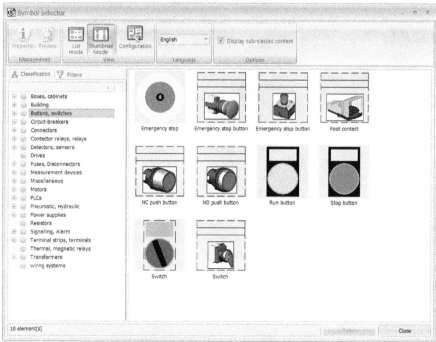

Figure-6. Symbol selector dialog box

- Select the desired category and then select the symbol you want to insert.
- Click on the **Select** button from the dialog box. The symbol will get attached to the cursor and the **Symbol insertion** options will be displayed in the **Command PropertyManager**; refer to Figure-7.

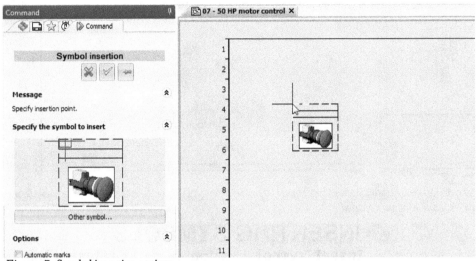

Figure-7. Symbol insertion options

- Click at the desired location in the drawing to specify the insertion point. The **Symbol properties** dialog box will be displayed; refer to Figure-8.

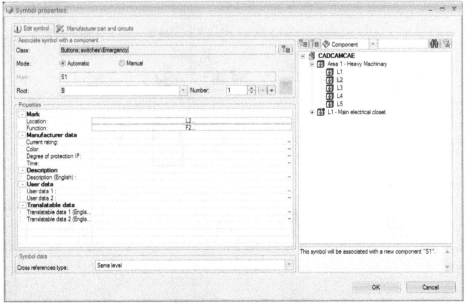

Figure-8. Symbol properties dialog box

- Specify the desired manufacturer data in the fields in **Properties** box of the dialog box.

Manufacturer parts and Circuits

- To select component data from the SolidWorks Electrical library, click on the **Manufacturer part and circuits** tab. The dialog box will be displayed as shown in Figure-9.

Figure-9. Symbol properties dialog box with manufacturer part and circuits

- Click on the **Search** button from the dialog box. The **Manufacturer part selection** dialog box will be displayed; refer to Figure-10.

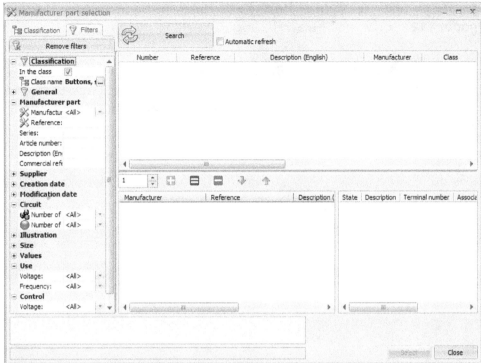

Figure-10. Manufacturer part selection dialog box

- Click on the ellipse button for **Class name** field under **Classification node** in the left of the dialog box; refer to Figure-11. The **Class selector** dialog box will be displayed; refer to Figure-12.

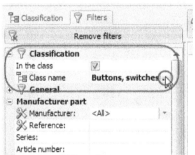

Figure-11. Ellipse button

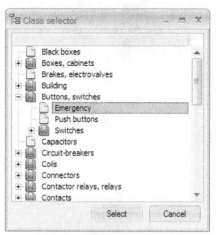

Figure-12. Class selector dialog box

- Select the class of component from the dialog box and click on the **Select** button.
- Click on the down arrow in the **Manufacturer** field of the **Manufacturer part** node and select the manufacturer of component; refer to Figure-13.

Figure-13. Manufacturer field

- Specify the other filters as required and click on the **Search** button from the dialog box. The related manufacturer parts will be displayed; refer to Figure-14.

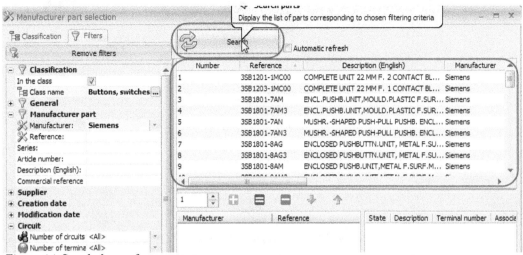

Figure-14. Searched manufacturer parts

- Select the desired component from the list and click on the **Add manufacturer part** button 🔲 from the dialog box. The selected component will be added in the current project; refer to Figure-15.

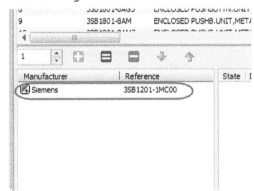

Figure-15. Part added in the project

- Similarly, you can add more manufacturer parts for the current project. Select the component from the new list of components added in the project and click on the **Select** button from the **Manufacturer part selection** dialog box. The **Symbol properties** dialog box will be displayed again.
- Select the desired component manufacturing description from list displayed in the **Manufacturer part and circuits** tab of the dialog box and click on the **OK** button from the dialog box. The component will be placed with component description; refer to Figure-16.

Figure-16. Component placed

If the **Symbols** palette is not displayed, then click on the **Symbols palette** button from the **View** panel in the **View** tab of the **Ribbon**; refer to Figure-17.

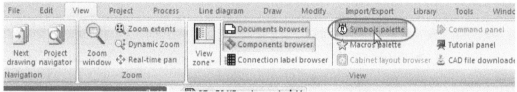

Figure-17. Symbols palette button

CONNECTION LABELS

A connection label is representation of a device in terms of connections. When we insert symbols by using the **Symbols** palette, we do not show the number of terminals of the part. To show the number of terminals along with the component symbol, we use the connection labels. A component which does not have a manufacturer part associated with it cannot be represented by a connection label. The procedure to insert a connection label is given next.

• Click on the **Insert a connection label for component** button from the **Connection label** drop-down in the **Insertion** panel of the **Ribbon**; refer to Figure-18. The **Command panel** will be displayed with options related to labels. Also, the component label will be attached to cursor; refer to Figure-19.

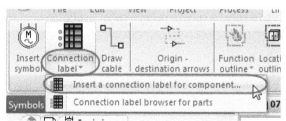

Figure-18. Insert a connection label for component button

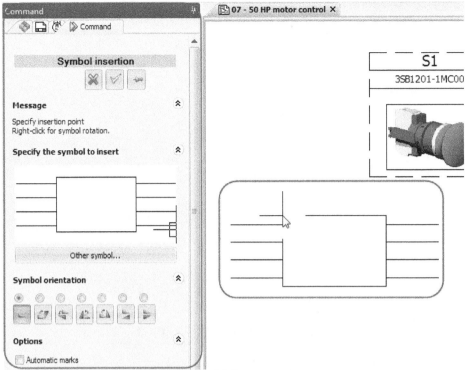

Figure-19. Command panel for connection label

- Click on the **Other symbol** button from the **Command panel** to select other symbol. The **Symbol selector** dialog box will be displayed; refer to Figure-20.
- Select the desired symbol from list and click on the **Select** button from the dialog box. The selected component will get attached to the cursor.
- Select the required radio button from the **Symbol orientation** rollout in the panel to rotate the symbol; refer to Figure-21.

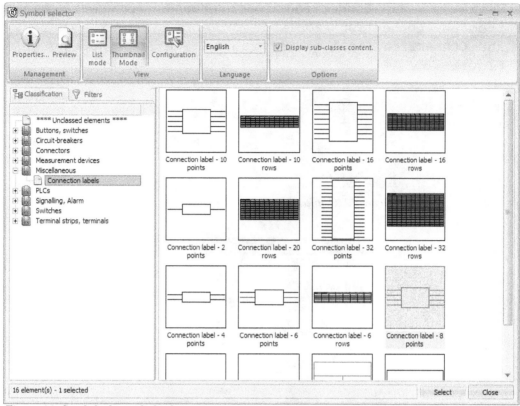

Figure-20. Symbol selector dialog box

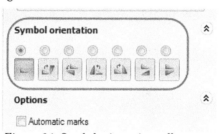

Figure-21. Symbol orientation rollout

- Click to place the symbol. The **Symbol properties** dialog box will be displayed as discussed earlier.
- Specify the desired attributes to the symbol and click on the **OK** button.

Connection label browser for parts

Earlier, we have learned to insert components in the project by using the Symbol palette. If you want to insert connection labels for the components earlier added in project then there is a direct method for that. The method is given next.

• Click on the **Connection label browser for parts** button from the **Connection labels** drop-down; refer to Figure-22. The **Connection label browser** will be displayed; refer to Figure-23.

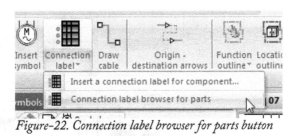

Figure-22. Connection label browser for parts button

Figure-23. Connection label browser

• Click on the check box for desired component in the Connection label browser, the label for component will get attached to cursor.
• Click in the drawing to place it; refer to Figure-24.

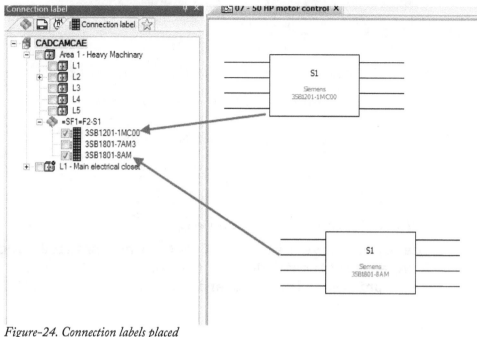

Figure-24. Connection labels placed

INSERTING PRINTED CIRCUIT BOARD

Sometimes in your electrical project, you may need to insert PCB (Printed Circuit Boards) to create desired functionality. In SolidWorks Electrical, you can insert PCB by using the **Insert printed circuit board** tool. The procedure to use this tool is given next.

- Click on the **Insert printed circuit board** tool from the **Insertion** panel in the **Line diagram** tab of the **Ribbon**. The **Printed circuit board insertion** dialog box will be displayed; refer to Figure-25.
- Click on the **Create a new printed circuit board from a manufacturer part** button if you have printed circuit boards in SolidWorks Electrical library and want to use them. Click on the **Create a new printed circuit board from an electronic design file** button if you want to use a electronic design file created by electronic design software.

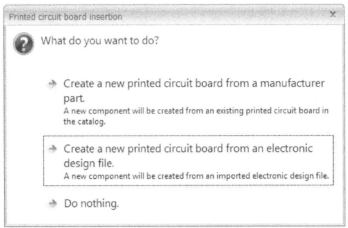

Figure-25. Printed circuit board insertion dialog box

- In our case, we have selected **Create a new printed circuit board from an electronic design file** button. On doing so, the **Select printed circuit board file** dialog box will be displayed; refer to Figure-26.

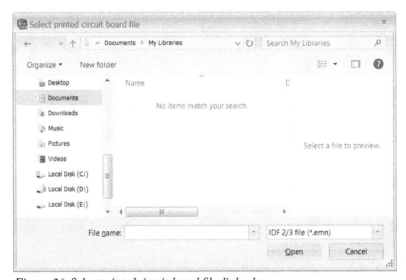

Figure-26. Select printed circuit board file dialog box

- Select the circuit file with **.emn** extension and click on the **Open** button from the dialog box. A SolidWorks message box will be displayed asking you whether to copy the file or let the system use original file.
- Click on the desired button from the message box. **Manufacturer Part properties** dialog box will be displayed; refer to Figure-27.

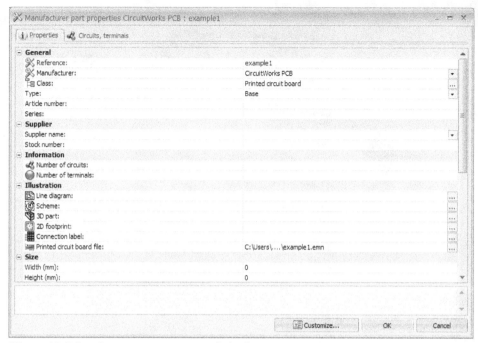

Figure-27. Manufacturer part properties dialog box

- Set the desired parameters and click on the **OK** button from the dialog box (You will learn about the options in this dialog box later in this book). The **Component properties** dialog box will be displayed (You will learn about options in this dialog box later in the book).
- Set the desired parameters and click on the **OK** button. **Symbol selector** dialog box will be displayed and you will be asked to select a symbol for circuit board.
- Double-click on the desired symbol in the dialog box. The symbol will get attached to cursor.
- Click at desired location to place the symbol.

DRAWING CABLES

Once you have inserted the desired components, the next step is to connect them with the help of a cable. The procedure to draw cable is given next.

- Click on the **Draw Cable** button from the **Insertion** panel in the **Line diagram** tab of the **Ribbon**. The **Command** panel will be displayed with the options related to cable; refer to Figure-28.

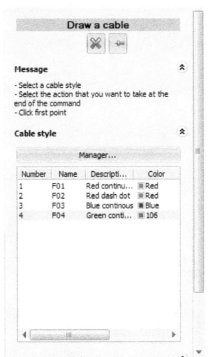

Figure-28. Command panel for cable

- Select the desired wire number from the list in the **Command** panel. You can also create a wire with desired properties by using the **Wire style manager** by clicking on the **Manager** button; refer to Figure-29.

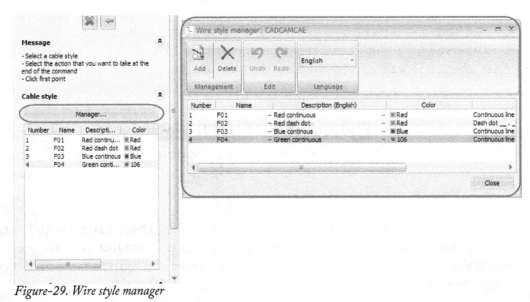

Figure-29. Wire style manager

- Click on the **Add** button from the **Wire style manager** and add the wire style with desired color and description.
- Click on the **Close** button from the **Wire style manager** and click on the component boundary to specify starting point of the wire; refer to Figure-30.

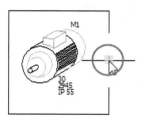

Figure-30. Starting point of wire

- Move the cursor and click at the desired location to make bend in the wire; refer to Figure-31.

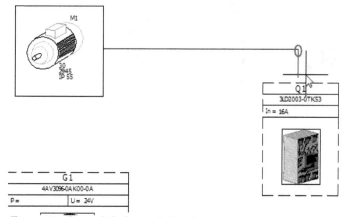

Figure-31. Point clicked to make bend

- Move the cursor upward/downward and click to specify the end point of the wire connected to the component. The **Cable insertion** dialog box will be displayed; refer to Figure-32.

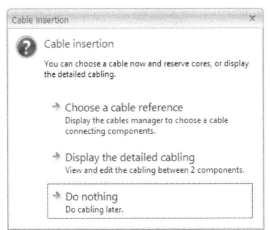

Figure-32. Cable insertion dialog box

- We will discuss about the **Display the detailed cabling** option later in this chapter. Click on the **Choose a cable reference** option from the dialog box. The **Cable references selection** dialog box will be displayed; refer to Figure-33.

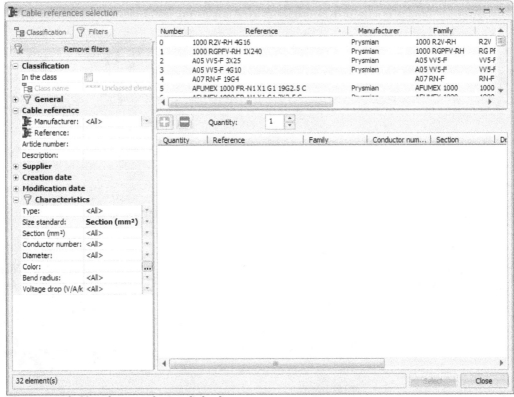

Figure-33. Cable references selection dialog box

- Select the desired cable from the list displayed at the top of the dialog box and click on the Add button; refer to Figure-34.

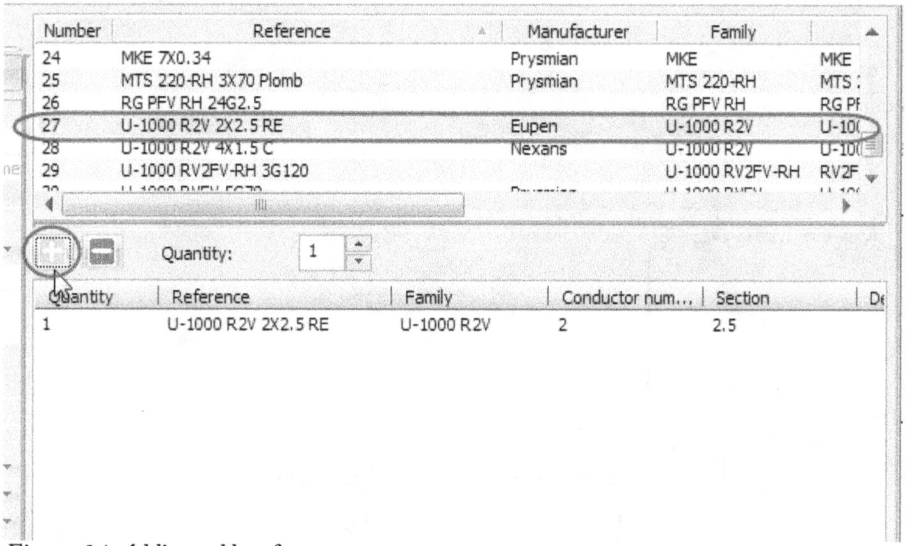

Number	Reference	▲	Manufacturer	Family	
24	MKE 7X0.34		Prysmian	MKE	MKE
25	MTS 220-RH 3X70 Plomb		Prysmian	MTS 220-RH	MTS :
26	RG PFV RH 24G2.5			RG PFV RH	RG PF
27	U-1000 R2V 2X2.5 RE		Eupen	U-1000 R2V	U-100
28	U-1000 R2V 4X1.5 C		Nexans	U-1000 R2V	U-100
29	U-1000 RV2FV-RH 3G120			U-1000 RV2FV-RH	RV2F
30	U-1000 RVFV FG70		Prysmian	U-1000 RVFV	U-100

Quantity: 1

Quantity	Reference	Family	Conductor num...	Section	De
1	U-1000 R2V 2X2.5 RE	U-1000 R2V	2	2.5	

Figure-34. Adding cable reference

- Select the cable reference from the list of added cable references and click on the **Select** button from the dialog box. The reference will be attached to the cable drawn; refer to Figure-35.

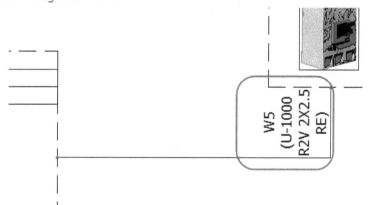

Figure-35. Cable reference attached

ORIGIN - DESTINATION ARROWS

The origin-destination arrows are used to link the components sharing same cable or wire but are in different drawings. The procedure to add origin-destination arrows is given next.

- Click on the **Origin - destination arrows** tool from the **Insertion** panel in the **Line diagram** tab of the **Ribbon**. The Origin-destination manager will be displayed; refer to Figure-36.

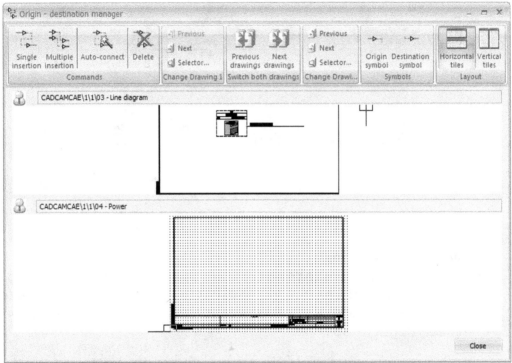

Figure-36. Origin-destination manager

- Using the options in the **Change Drawing** panels, change the drawing being displayed if you want the different then being displayed; refer to Figure-37.

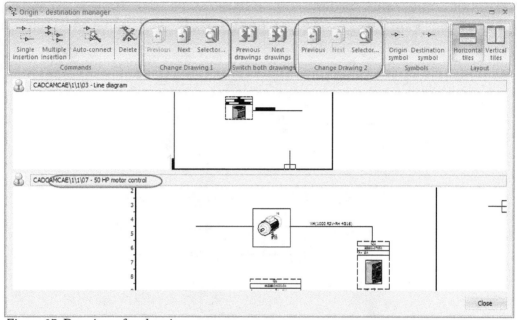

Figure-37. Drawings after changing

- Click on the **Auto-connect** button from the **Commands** panel if the cable are of same number.
- To manually set the link, click on the **Single insertion** button from the **Commands** panel in the **Ribbon** of dialog box. Open end will automatically get highlighted in the drawing; refer to Figure-38.

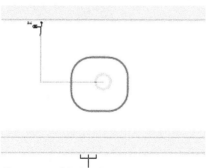

Figure-38. Highlighted open end

- Click on the end highlighted by green circle in the first drawing. It will turn into red circle and you are asked to selected the other end point.

- Select the end point of cable/wire in other drawing; refer to Figure-39. Arrow heads will get attached to the cursor; refer to Figure-40.

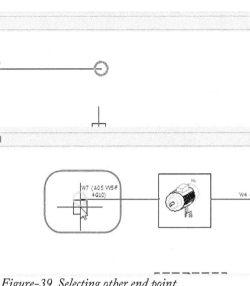

Figure-39. Selecting other end point

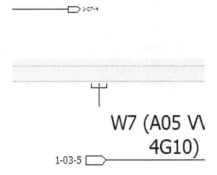

Figure-40. Arrow heads attached

- To change the arrow head style, click on the **Origin Symbol** or **Destination Symbol** button (as required) from the **Symbols** panel in the **Origin-destination manager**. The respective dialog box will be displayed; refer to Figure-41.

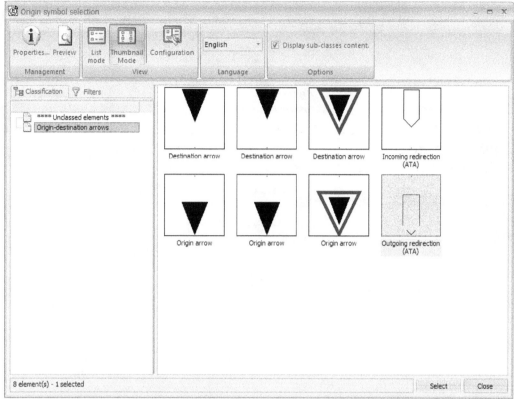

Figure-41. Origin symbol selection dialog box

- Select the desired style from the templates and click on the **Select** button.
- In the same way, you can connect other cables/wires in different drawings.

FUNCTION OUTLINE

The Function outlines are used to mark the components on the basis of their functions in the circuit. These outlines are just to categorize the components in drawing on the basis of function. The procedure to create function outlines is given next.

- Click on the down arrow below **Function outline** button in the **Insertion** panel of the **Line diagram** tab in the **Ribbon**. The tools for outlining will be displayed; refer to Figure-42.

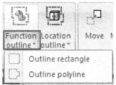

Figure-42. Tools for function outlining

- Select the desired tool from the list displayed (**Outline polyline** selected in our case) and drawing a boundary for the components for same function; refer to Figure-43.

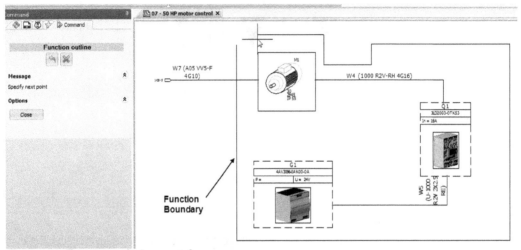

Figure-43. Function outlining created

- Click on the **Close** button from the **Command** panel to close the polyline boundary. The **Select function** dialog box will be displayed; refer to Figure-44.

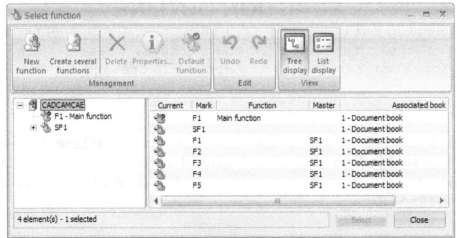

Figure-44. Select function dialog box

- Select the desired function from the list and click on the **Select** button from the dialog box. The **Change component function** dialog box will be displayed; refer to Figure-45.

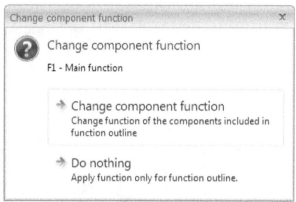

Figure-45. Change component function dialog box

- Click on the **Change component function** button from the dialog box to change the function of the components enclosed in the boundary.

LOCATION OUTLINE

The Location outlines are used to mark the component on the basis of their locations in the circuit. The procedure to create location outline is same as for the Function outline.

DETAILED CABLING

Detailed cabling is used to represent the connections of cable with various components. In other words, we can describe the connection between various components of circuit with the help of detailed cabling. The procedure to use detailed cabling is given next.

- Click on the **Detailed Cabling** tool from the **Changes** panel in the **Line diagram** tab of the **Ribbon**. The **Detailed cabling** command panel will be displayed at the left and your are asked to select a cable.

- Click on the desired cable for which you want to specify the detailed connections and press ENTER from keyboard. The **Detailed cabling** dialog box will be displayed; refer to Figure-46.

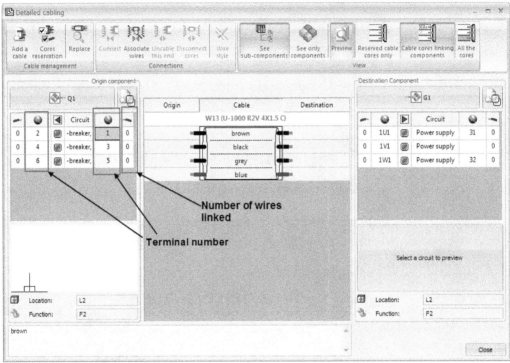

Figure-46. Detailed cabling

The dialog box is divided into three sections; Origin component, Cable and Destination component; refer to Figure-47. To connect the origin component, use the options in **Origin component** section and similarly for the **Destination component**. Procedure is given next.

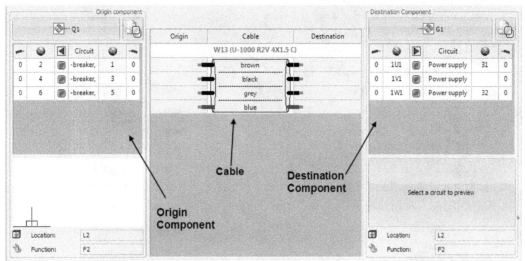

Figure-47. Sections of Detailed cabling dialog box

- Click in the wire box next to terminal 1 in the Origin component section and then click on the left green box for brown wire; refer to Figure-48.

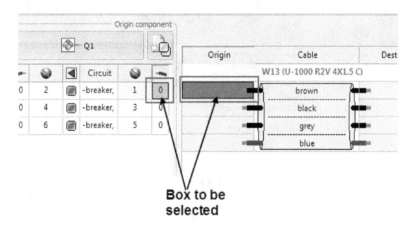

Figure-48. Boxes selected for connection

- Click on the **Connect** button from the **Connections** panel in the dialog box. The connection will be created as shown in Figure-49. Here, **Q1** is name of origin component and **1** is the terminal number in **Q1:1**.

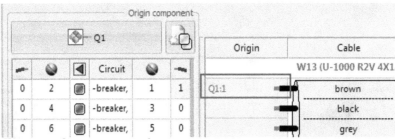

Figure-49. Connection created

- Similarly, connect the other side of brown wire to the 1U1 terminal of destination component and repeat the process for other wires in the cable; refer to Figure-50.

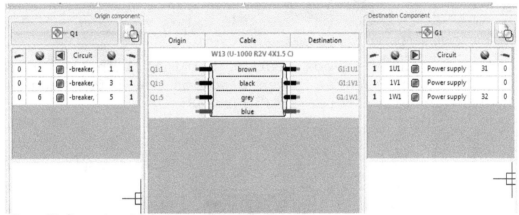

Figure-50. Connection with destination component

- Now, we are left with blue wire and we don't have terminals to connect it with. We are going to use this wire ground. To make a terminal for ground in the Origin component, click on the **Add virtual circuits** button from the **Origin component** section; refer to Figure-51. The **Add virtual circuits to component** dialog box will be displayed; refer to Figure-52.

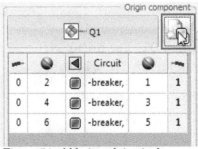

Figure-51. Add virtual circuits button

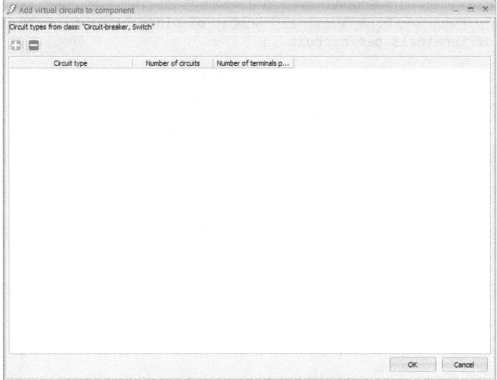

Figure-52. Add virtual circuits to component dialog box

- Click on the plus button from the dialog box. A new circuit will be added in the list. Click on the **More circuit types** option from the drop-down list displayed on clicking down arrow in **Circuit:type** column; refer to Figure-53.

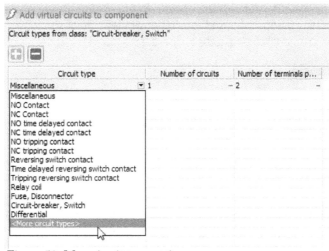

Figure-53. More circuit types option

- Click on the **Ground** option from the drop-down list; refer to Figure-54. Specify the **Number of circuits** as **1** and **Number of terminals per circuit** as **2**.

Figure-54. Ground option

- Click on the **OK** button from the dialog box.
- Connect the ground with the blue wire and do the same procedure on other side; refer to Figure-55.

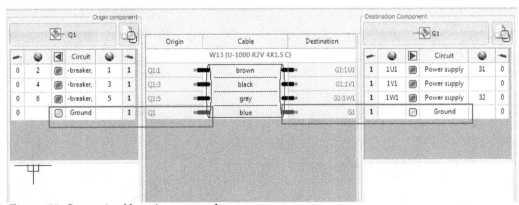

Figure-55. Connecting blue wire to grounds

- To disconnect cable or cores, click on the **Uncable this end** or **Disconnect cores** button from the dialog box, respectively.

FOR STUDENT NOTES

Chapter 4

Schematic Drawing

Topics Covered

The major topics covered in this chapter are:

- *Introduction*
- *Starting Schematic Drawings*
- *Inserting Symbols, Wires, and Black box*
- *Inserting PLCs, Connectors, and Terminals*
- *Inserting Reports*

INTRODUCTION

As discussed earlier in the book, schematic drawing is the drawing showing all significant components and parts of a circuit with their interconnections. The tools to create schematic drawing are available in the **Schematic** tab of **Ribbon**; refer to Figure-1. Note that this tab will be available only when you have started schematic drawing to work upon. The procedure to start a schematic drawing in project is given next.

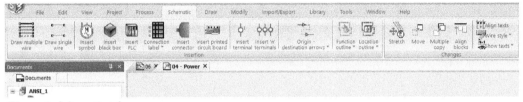

Figure-1. Schematic tab

STARTING A SCHEMATIC DRAWING

- Click on the **New scheme** tool from the **New** drop-down in the **Project** panel of the **Project** tab in the **Ribbon**; refer to Figure-2. A new drawing will be added to the document book.
- Right-click on the newly added drawing in the **Documents browser** and select the **Properties** option from the shortcut menu. The **Drawing** dialog box will be displayed; refer to Figure-3.

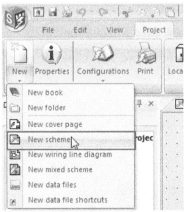

Figure-2. New scheme tool

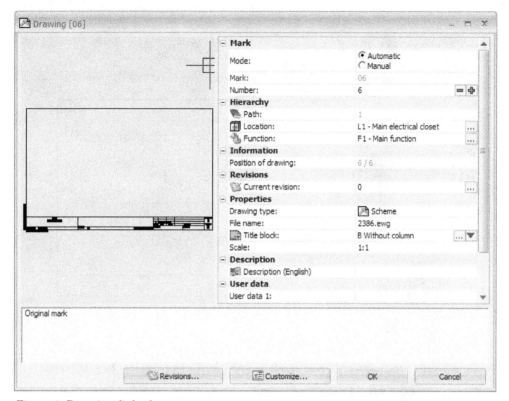

Figure-3. Drawing dialog box

- Click in the **Description** field and specify the description for drawing.
- Set the location and function of the drawing and click on the **OK** button from the dialog box to start the drawing. The drawing will be modified accordingly.
- Double-click on the drawing file name in the **Documents browser** to open it if not opened earlier. The tools related to schematic drawing will be displayed; refer to Figure-4.

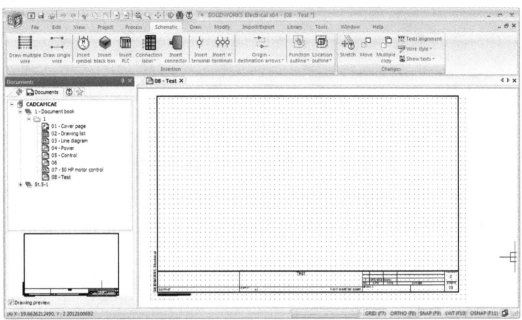
Figure-4. Starting schematic drawing

INSERTING SYMBOL

The procedure to insert symbols in schematic drawing is similar to the procedure of inserting symbol in line diagrams. The procedure to insert schematic symbols is given next.

- Click on the **Insert symbol** tool from the **Insertion** panel in the **Schematic** tab of the **Ribbon**. If the **Symbol insertion** page will be displayed in the **Command** panel; refer to Figure-5.
- Click on the **Other symbol** button from the **Command panel**. The **Symbol selector** dialog box will be displayed; refer to Figure-6.

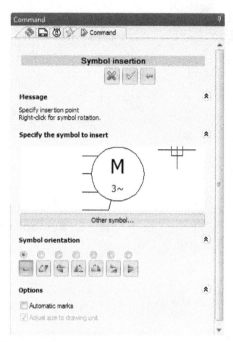

Figure-5. Symbol insertion page in Command panel

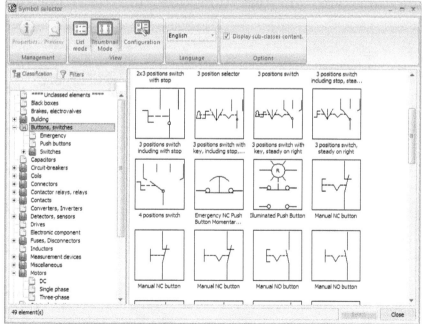

Figure-6. Symbol selector dialog box

- Select the desired symbol from the dialog box and click on the **Select** button. The symbol will get attached to cursor.

- Select the desired orientation for the symbol from the **Symbol orientation** area in the **Command** panel; refer to Figure-7.

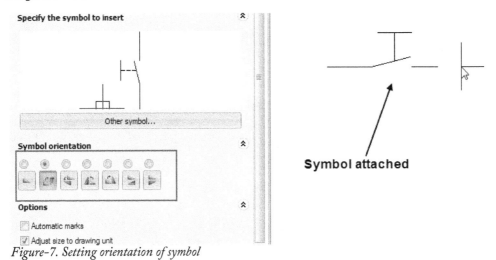

Figure-7. Setting orientation of symbol

- Click in the drawing to place the symbol. The **Symbol properties** dialog box will be displayed; refer to Figure-8.

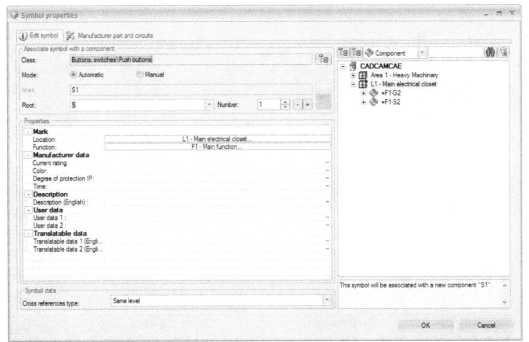

Figure-8. Symbol properties dialog box

- Click on the **Manufacturer part and circuits** tab and set the manufacturer data for component. The options in the dialog box have been discussed earlier in the book.
- Click on the **OK** button from the dialog box. The symbol will be displayed along with its attributes; refer to Figure-9.

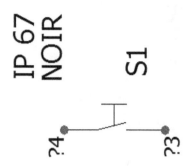

Figure-9. Symbol with attributes

INSERTING WIRES

We have wires for schematic drawings like, cables for Line diagrams. We have two tools named **Draw multiple wire** and **Draw single wire** to insert wires in the schematic drawings. The procedures to create both the wires are given next.

Inserting Single Wire

Wires are used to connect the components so that they can function as required. The procedure to insert single wire in the drawing is discussed next.

- Click on the **Draw single wire** tool from the **Insertion** panel of the **Schematic** tab in the **Ribbon**. The **Electrical wires** page will be displayed in the **Command** panel; refer to Figure-10.

Figure-10. Electrical wires page in Command panel

- Click on the ellipse button [...] next to **Name** field in the **Wire style selection** area of the **Command** panel. The **Wire style selector** dialog box will be displayed; refer to Figure-11.

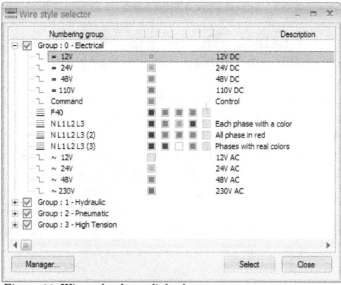

Figure-11. Wire style selector dialog box

- Select the desired wire from the list in the dialog box.
- If the desired wire is not available in the list, click on the **Manager** button from the dialog box. The **Wire style manager** dialog box will be displayed; refer to Figure-12.

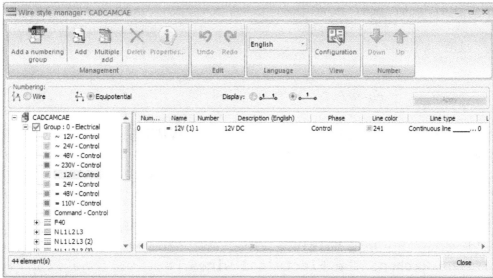

Figure-12. Wire style manager

- Create a wire style with desired parameters as discussed in previous chapter. Select the newly created wire style.
- Click on the **Select** button from the **Wire style selector** dialog box. You are asked to specify the start point of the wire.
- Click to specify the starting point. You are asked to specify the next point of the wire. Press **F8** if you want to create non-ortho wire.
- Click consecutively to specify the corners of wire and press **ENTER** when you want to exit the wire creation.
- If you want to create more than one wires then set the desired number in the **Number of lines** spinner in the **Electrical wires** page of **Command** panel; refer to Figure-13. Rest of the procedure is same.

Figure-13. Number of lines spinner

Inserting Multiple Wires

For electrical supplies like three phase connection, we need set of four wires. Such connections can be made by using the **Draw multiple wire** tool. The procedure to use this tool is given next.

- Click on the **Draw multiple wire** tool from the **Insertion** panel of the **Ribbon**. The **Electrical wires** page will be displayed in the Command panel. If you are using this tool for the first time, then the **Wire style selector** dialog box will be displayed. Select the desired wire from the dialog box. Note that the wire with the multiple wire icon are used to draw multiple wires.

- Select the desired check box from the **Available wires area** to enable wires in the wire set; refer to Figure-14.

Figure-14. Available wires area

- You can specify the distance between two consecutive wire lines by using the **Space between lines** edit box.
- After setting the desired parameters, click in the drawing area to specify the starting point for the wires. The procedure of drawing wire is same as for single wire.

INSERTING BLACK BOX

Black box is a customizing object in SolidWorks electrical. We can use black box when we don't want or say we don't have the symbol for our component. Black box has a specify property that when you place a black box on any wire or set of wires, the respective number of terminals are automatically created on it. The procedure to insert black box in drawing is given next.

- Click on the **Insert black box** tool from the **Insertion** panel in the **Schematic** tab of the **Ribbon**. If you have selected a symbol of black box earlier, then the **Symbol insertion** page will be displayed in the **Command** panel as shown in Figure-15.

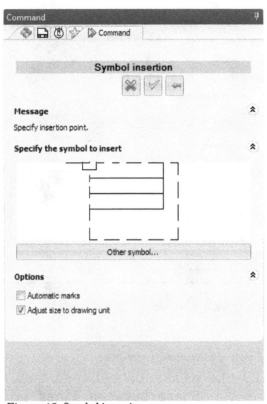

Figure-15. Symbol insertion page

- If you are using this tool for the first time then the **Symbol selector** dialog box will be displayed; refer to Figure-16.

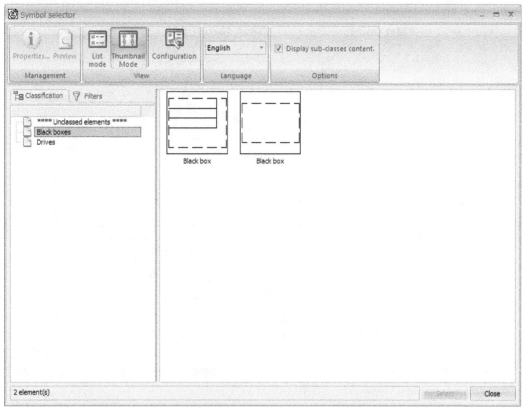

Figure-16. Symbol selector dialog box for black box

- Select the desired symbol and click on the **Select** button from the dialog box.
- Click in the drawing to specify the starting point of black box. You are asked to specify the other corner point for rectangular boundary of black box; refer to Figure-17.

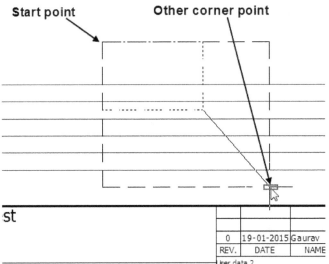

Figure-17. Creating boundary of black box

- Click to specify the other corner point in such a way that the desired wires (which you want to connect with black box) are overlapped by the rectangle. On specifying the corner point, the **Symbol properties** dialog box will be displayed as discussed earlier.
- Set the desired properties in the dialog box and click on the **OK** button from the dialog box. The symbol will be connected to the covered wires; refer to Figure-18.

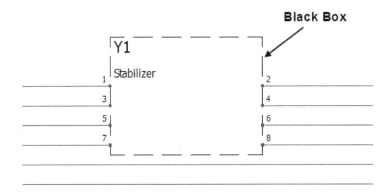

Figure-18. Black box symbol placed

INSERTING PLC

PLC is a solid state/ computerized industrial computer that performs discrete or sequential logic in a factory environment. It was originally developed to replace mechanical relays, timers, counters. PLCs are used successfully to execute complicated control operations in a plant. Its purpose is to monitor crucial process parameters and adjust process operations accordingly. A sequence of instructions is programmed by the user to the PLC memory and when the program is executed, the controller operates a system to the correct operating specifications.

PLC consists of three main parts: CPU, memory and I/O units.

CPU is the brain of PLC. It reads the input values from inputs, runs the program existed in the program memory and writes the output values to the output register. Memory is used to store different types of information in the binary structure form. The memory range of S7-200 is composed of three main parts as program, parameter, and retentive data fields. I/O units provide communication between PLC control systems.

The procedure to insert PLC in drawing is discussed next.

• Click on the **Insert PLC** tool from the **Insertion** panel of the **Schematic** tab in the **Ribbon**. The **Manufacturer part selection** dialog box will be displayed; refer to Figure-19.

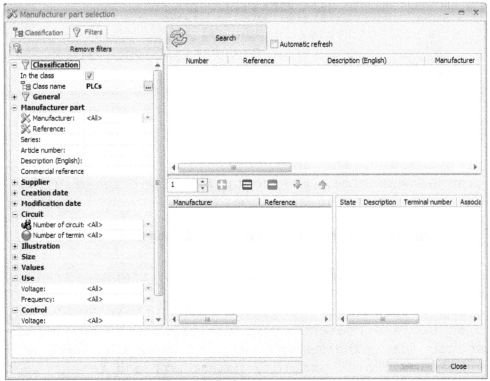

Figure-19. Manufacturer part selection dialog box

- Search the PLC with desired number of terminals and properties and click on the **+** button to add it in project.
- Click on the **Select** button to select it. The **Component properties** dialog box will be displayed.
- Specify the user data in the dialog box and click on the **OK** button. PLC will get attached to cursor and parameters related to channels will be displayed in the **Command** panel; refer to Figure-20.

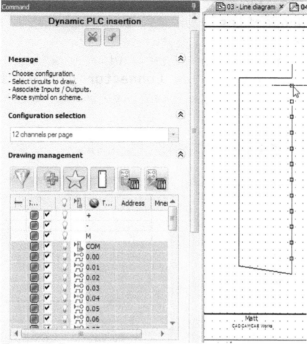

Figure-20. Parameters related to channels

- Select the check boxes from the **Channel selection** area in the **Command panel** to enable the respective connection points.
- Similarly, set the other parameters and click in the drawing to place the PLC.

You can insert Printed Circuit Board as discussed in previous chapter.

INSERTING CONNECTORS

Connectors are used to facilitate easy assembly of components to the circuit. The most common example of connectors can be the power outlet and power plug. There is a long list of connectors available in the market so we are not going to discuss details of them in this book. The procedure to insert all the types of connectors is same and is given next.

- Click on the **Insert Connector** tool from the **Insertion** panel in the **Schematic** tab of the **Ribbon**. The **Manufacturer part selection** dialog box will be displayed.
- Search the connector with desired number of terminals by using filters and add it to the project.

- Click on the **Select** button from the dialog box. The **Component properties** dialog box will be displayed as discussed earlier.
- Set the desired parameters and click on the **OK** button from the dialog box. The **Connector dynamic insertion** page of **Command** panel will be displayed and connector will get attached to cursor; refer to Figure-21.

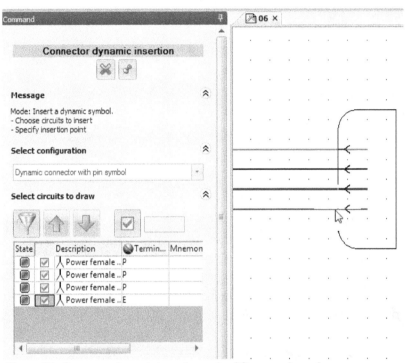

Figure-21. Connector dynamic insertion page of Command panel

- Select the desired option from the **Select configuration** drop-down in the panel. If you select the **Dynamic connector with pin symbol** option then connector will be displayed with pins; refer to Figure-22. If you select the **Dynamic connector without pin symbol** option then connector will be displayed without pins; refer to Figure-22. If you have selected the **One symbol per pin** option then connector symbol will not be displayed but pin symbol will be applied to every selected connection; refer to Figure-22. If you want to insert a symbol available in the **Symbol selector** dialog box then select the **<Select symbol to insert>** option from the drop-down and click on the **Other symbol** button from the panel. The **Symbol selector** dialog box will be displayed.

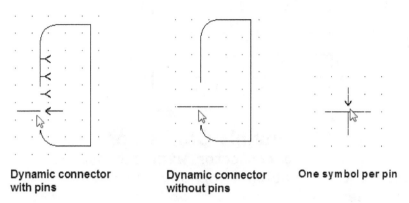

Dynamic connector with pins   **Dynamic connector without pins**   **One symbol per pin**

Figure-22. Connector options

- Select the desired symbol and click on the **OK** button from the dialog box. The symbol will get attached to the cursor; refer to Figure-23.

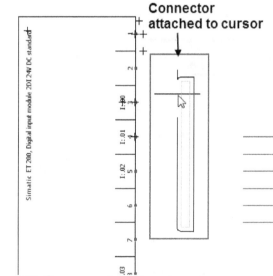

Figure-23. Connector attached to cursor

- Hover the cursor to a terminal till you get a snap point and then click, the connector will automatically create connections with the terminal. Refer to Figure-24 in which wires are connected to the connector.

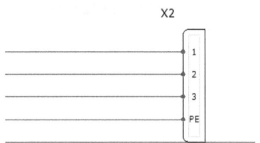

Figure-24. Wires connected to connector

Dynamic Connector

- Select the **Dynamic connector with pin symbol** option from the **Select configuration** drop-down. You will be asked to specify the insertion point of the connector.
- Click on the wire/terminal that you want to connect. The connector will be created; refer to Figure-25.

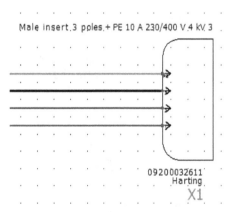

Figure-25. Connector created by dynamic connector option

You will learn about customizing connector in the next chapter.

INSERTING TERMINAL /TERMINALS

Terminals are used to allow connection of wires to the main circuit. In other words, terminals allow branch circuits to be connected with the main circuit. The procedure to insert terminal /terminals is given next.

Inserting single terminal

- Click on the **Insert terminal** button from the **Insertion** panel in the **Ribbon**. The **Terminal selector** dialog box will be displayed (for the first time users); refer to Figure-26.

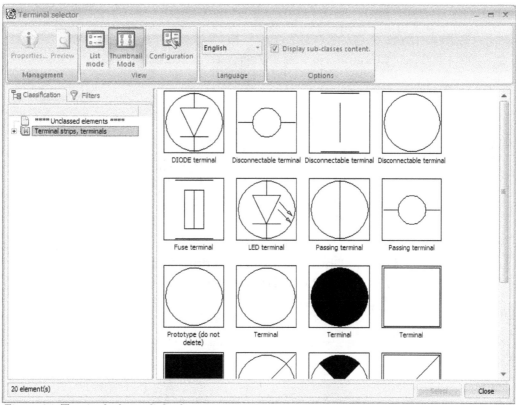

Figure-26. Terminal selector dialog box

- Select the desired terminal symbol and click on the **Select** button from the dialog box. The symbol will get attached to the cursor.
- Click on the wire to insert terminal; refer to Figure-27. You are asked to specify the orientation of the terminal.

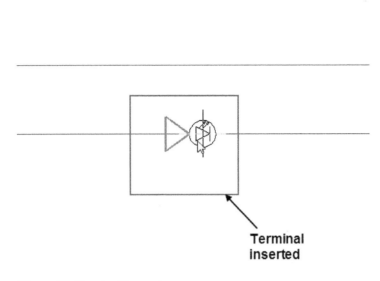

Figure-27. Terminal inserted

- Click on desired side of terminal to specify the orientation of the terminal. The **Terminal properties** dialog box will be displayed; refer to Figure-28.

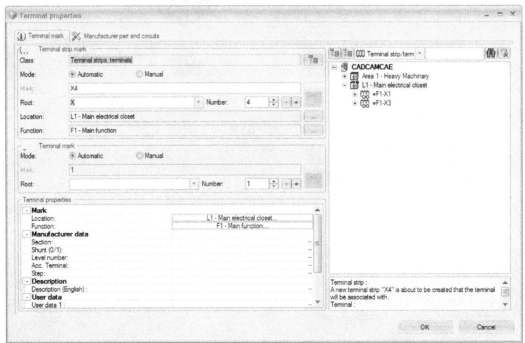

Figure-28. Terminal properties dialog box

- Specify the desired parameters and click on the **OK** button from the dialog box to create the terminal. If you want to associate a new terminal to terminal earlier created then follow the steps given in Figure-29 and click on the **OK** button.

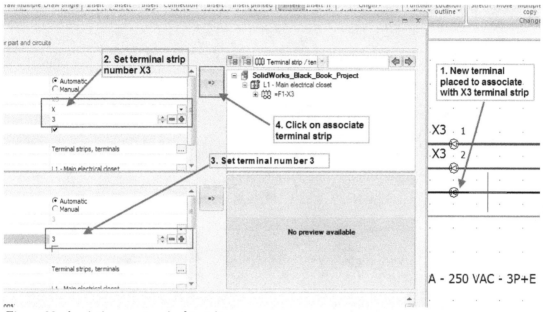

Figure-29. Associating new terminal to strip

Inserting multiple terminals

- Click on the **Insert 'n' terminals** button from the **Insertion** panel in the **Ribbon**. If you are using the tool for the first time then **Terminal selector** dialog box will be displayed. Double-click on the desired symbol from the **Terminal selector** dialog box. The **Terminal insertion** page will be displayed in the **Command** panel; refer to Figure-30 and you will be asked to draw an axis line intersecting with wires for creating terminals.

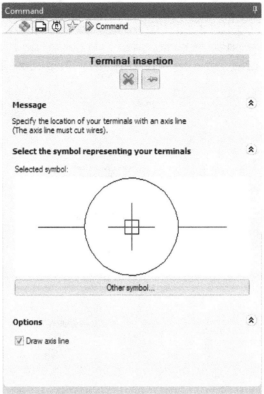

Figure-30. Terminal insertion page

- Click on the **Other symbol** button and select the symbol as discussed earlier if you want to change the terminal symbol.
- Draw an axis intersecting the wires to create terminals; refer to Figure-31.

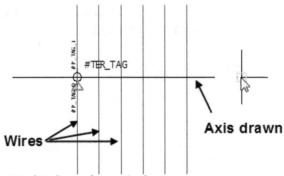

Figure-31. Axis drawn for terminals

- Click at the desired side of terminal symbol to set the orientation. The **Terminal symbol properties** dialog box will be displayed.

- Set the properties and click on the **OK (all terminals)** button to apply the same properties to all the terminals or you can apply individual properties by using the **OK** button. The terminals will be created in the form of a strip aligned to axis; refer to Figure-32.

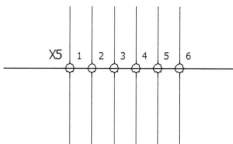

Figure-32. Terminal created

Till this point, we have covered almost all the tools related to schematics. The tools like **Origin-destination arrows**, **Function outline**, and **Location outline** have already been discussed in previous chapters. Now, we will discuss about editing terminal strips and inserting reports in the project.

TERMINAL STRIP EDITOR

Terminals are the connecting points used for various circuits so it is important to manage terminals properly. There is a special tool to manage connections of terminals at one place. After creating terminal, follow the procedure given next to manage terminal strip.

- Click on the **Terminal strips** tool from the **Management** panel in the **Project** tab of **Ribbon**. The **Terminal strips manager** dialog box will be displayed; refer to Figure-33.

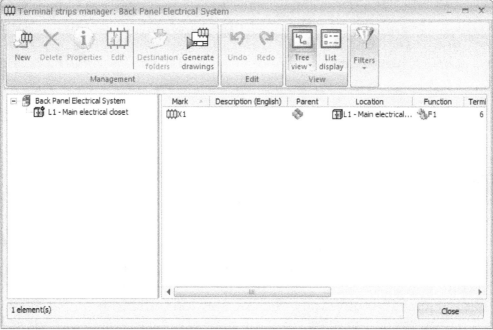

Figure-33. Terminal strips manager dialog box

- List of terminals used in the current project will be displayed in the dialog box. Double-click on the terminal strip from the dialog box. The **Terminal strip editor** dialog box will be displayed; refer to Figure-34.

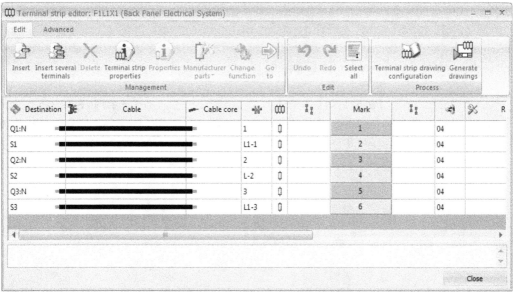

Figure-34. Terminal strip editor dialog box

- From this dialog box, we can find out how a component is connected to a terminal. From the above figure, we can find out that Neutral point of Circuit Breaker 1 is connected to first terminal of terminal strip. But, we cannot find out the type of cable used to this connection and manufacturer data for terminal. These parameters are specified in this dialog box.
- To specify the cable type, right-click in the cell under **Cable** column in the dialog box. A shortcut menu will be displayed; refer to Figure-35.

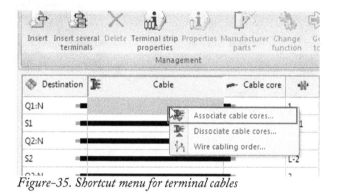

Figure-35. Shortcut menu for terminal cables

- Click on the **Associate cable cores** option from the shortcut menu. The **Associate cable cores** dialog box will be displayed; refer to Figure-36.

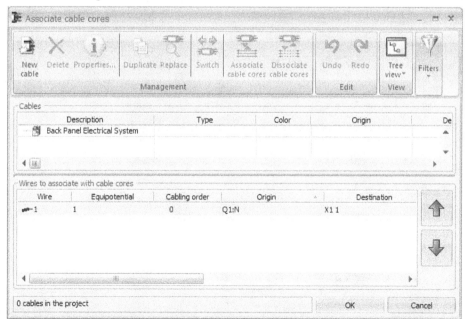

Figure-36. Associate cable cores dialog box

- Click on the **New cable** button from the **Management** panel in the dialog box. The **Cable references selection** dialog box will be displayed; refer to Figure-37.

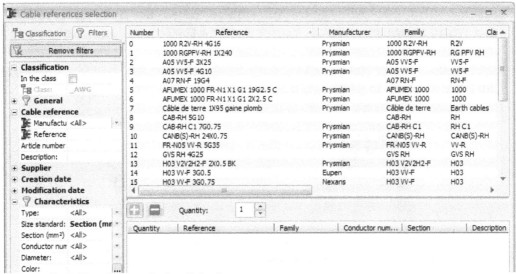

Figure-37. Cable references selection dialog box

- Double-click on the 10 AWG wire from Lapp manufacturer by using filters and click on the **Select** button. The wire will be added in the **Associate cable cores** dialog box.
- Select the wire and click on the **Associate cable cores** button from the dialog box; refer to Figure-38. The wire will be associated with selected cable.

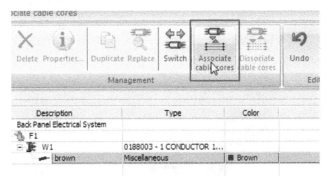

Figure-38. Associate cable cores button

- Similarly, you can associate other wires to cables. Note that if you need the same wire for next association then you can duplicate the cable by using **Duplicate** button after selecting cable from the **Associate cable cores** dialog box; refer to Figure-39.

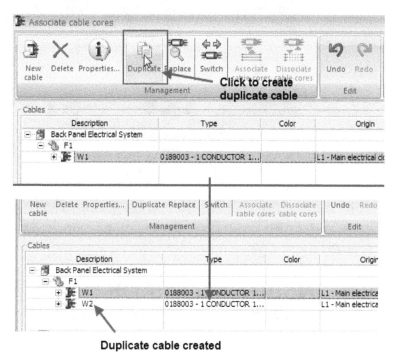

Figure-39. Creating duplicate cable

- To define manufacturer part for a terminal in terminal strip, click in the field under **Reference** column and click on the **Assign manufacturer parts** option from the **Manufacturer parts** drop-down in the **Terminal strip editor** dialog box; refer to Figure-40. The **Manufacturer part selection** dialog box will be displayed.

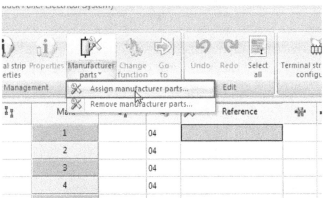

Figure-40. Assign manufacturer parts option

- Select the desired manufacturer part for the terminal. Similarly, select the manufacturer part for other terminals.

Note that if you have already associated cables in the schematic drawing then you do not need to specify the cable association again.

INSERTING REPORTS

SolidWorks Electrical has a dedicated tool for generating reports. Some of the reports that can be generated in SolidWorks Electrical are; cabling, wiring, Bill of Material, Drawing list, and so on. Procedure to insert reports in the project is given next.

- Click on the **Reports** button from the **Reports** panel in **Project** tab of the **Ribbon**; refer to Figure-41. The **Report manager** dialog box will be displayed as shown in Figure-42.

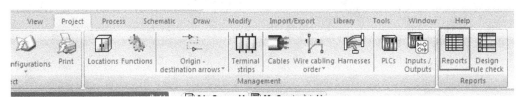

Figure-41. Reports tool

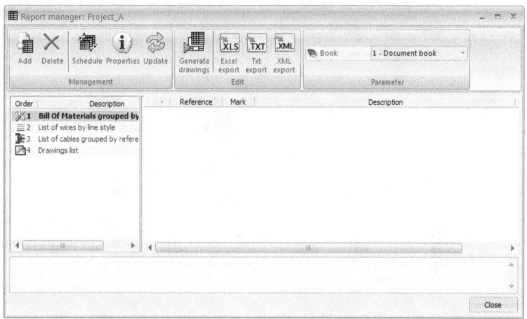

Figure-42. Report Manager

- By default, four reports are displayed in the left of the **Report manager** for Bill of Materials, list of wires, list of cables, and list of drawings. To add a new report, click on the **Add** button from the **Report manager**. The **Report configuration selector** dialog box will be displayed; refer to Figure-43. Select the check box for desired report and click on the **OK** button.
- To display the content of the report, click on it from the left of **Report manager**.

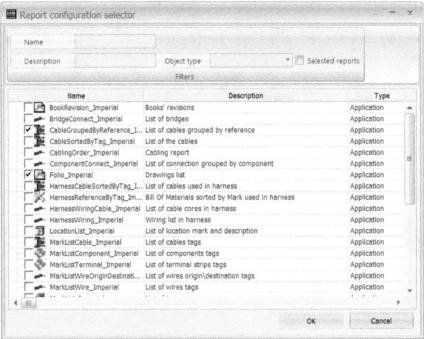

Figure-43. Report configuration selector dialog box

- To generate drawing from the report, click on the report from the left of **Report manager** and then click on the **Generate drawings** button from **Report manager**; refer to Figure-44. The drawing will be added in the project; refer to Figure-45.

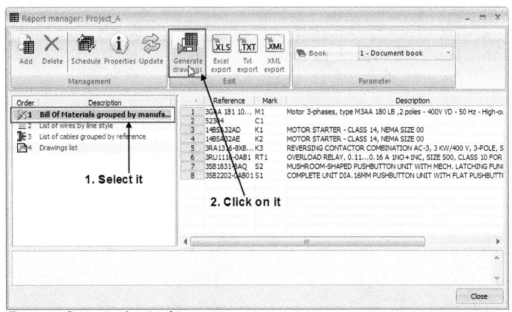

Figure-44. Generating drawing from report

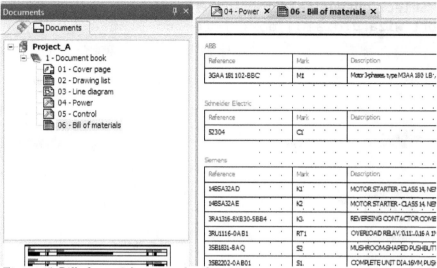

Figure-45. Bill of materials generated as drawing

- You can export the report to external formats like Excel, Txt, and XML by using **Excel export**, **Txt export**, and **XML export** tool, respectively. To do so, click on the respective button (Excel export in our case), the export wizard will be displayed; refer to Figure-46.

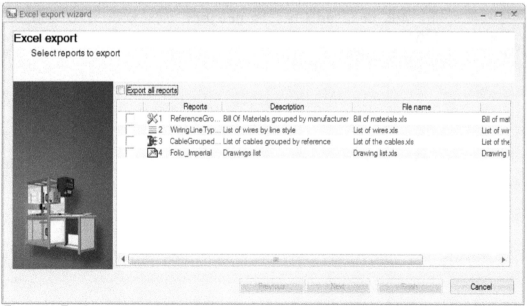

Figure-46. Excel export wizard dialog box

- Select check box/boxes from the list which you want to export and click on the **Next** button from the dialog box.

The **Select output folder** page will be displayed in the **Excel export wizard** dialog box; refer to Figure-47.

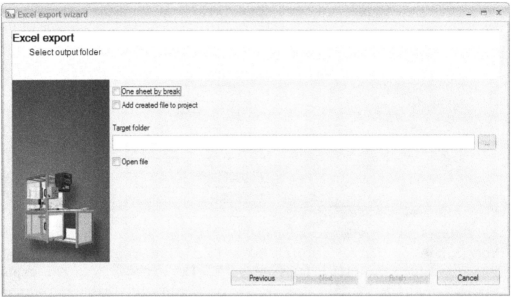

Figure-47. Select output folder page

- Select the **One sheet by break** check box if you want to write all the reports in one sheet separated by breaks.
- Select the **Add created files to project** check box to add the exported files into project also.
- Click on the **Finish** button to export files.

Chapter 5

Wire Numbering and Customizing

Topics Covered

The major topics covered in this chapter are:

- **Wire Numbering and Renumbering**
- **Harness**
- **Cable Management**
- **Wire Cabling Order**
- **PLCs Manager**
- **PLC Input/Output Manager**
- **Title Block Manager**
- **2D Footprint Manager**
- **Symbols Manager**
- **Cable references manager**
- **Manufacturer parts manager**

ADDING WIRE NUMBERS MANUALLY

Wiring numbers are used to identify the wires and their connections. The procedure to display wire numbers is given next.

- Click on the **Number new wires** tool from the **Processes** panel in the **Process** tab of the **Ribbon**. The **SOLIDWORKS Electrical** dialog box will be displayed.
- Click on the **Yes** button from the dialog box. The wiring numbers will be displayed; refer to Figure-1.

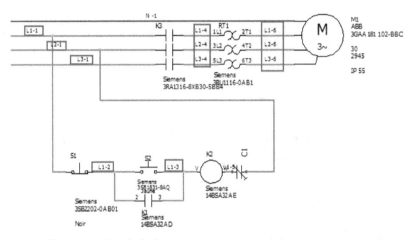

Figure-1. Wiring numbers linked to wires

- Click on the **Renumber wires** tool from the **Processes** panel in the **Process** tab of **Ribbon** if you have made changes in the wiring and want to reflect it in the circuit diagram.

Editing Wire Number Formula

By default, a number is assigned to the wire denoting its wire number but terminals of symbols also have numbers which can sometimes cause confusion. To solve this problem, we can set wire numbers to be displayed as **Wire n**. Here **n** is the number of wire. Let's see how we can do it.

- Click on the **Wire styles** option tool from the **Configurations** drop-down of the **Project** panel in the **Project** tab of the **Ribbon**. The **Wire style manager** will be displayed as discussed earlier.

- Select the wire whose wire numbering formula is to be changed and right-click on it. A shortcut menu will be displayed; refer to Figure-2.

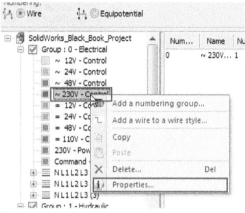

Figure-2. Right-click shortcut menu for wire

- Select the **Properties** option from the menu. The **Wire style** dialog box will be displayed; refer to Figure-3.

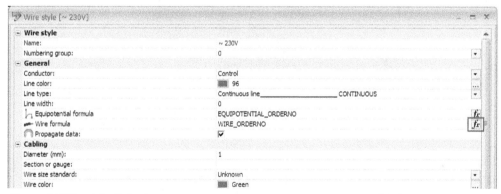

Figure-3. Wire style dialog box

- Click on the **fx** button next to **Wire formula** field in the **General** node of dialog box. The **Formula Manager** will be displayed; refer to Figure-4.
- Double-click on **Wire number** field if not selected by default. Click in the **Formula: Wire mark** field at the bottom in the dialog box and specify the formula as **"Wire"+ WIRE ORDERNO**; refer to Figure-5.

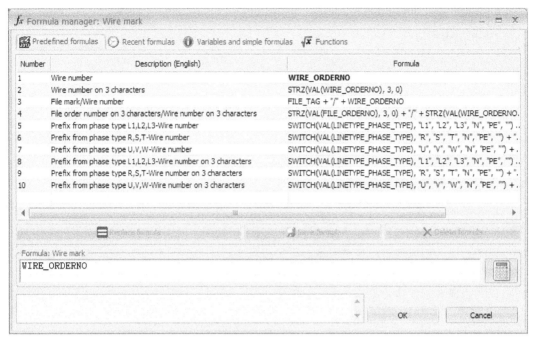

Figure-4. Formula manager1

Figure-5. Formula specified in the bottom field

- Click on the **OK** button from the dialog box and then click on the **OK** button from the **Wire style** dialog box. Close the **Wire style manager**. Now, check the wire numbering.

Renumbering Wires

- If it did not change then click on the **Renumber wires** tool from the **Processes** panel of the **Process** tab in the **Ribbon**. The **Renumber Wires** dialog box will be displayed; refer to Figure-6.

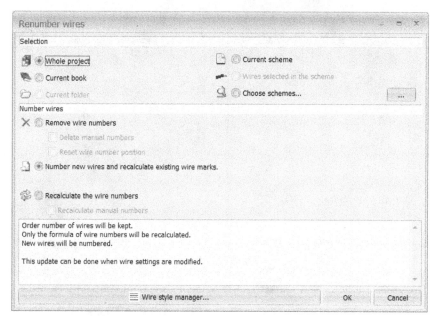

Figure-6. Renumber wires dialog box

- Select the desired radio button from **Whole project**, **Current book**, **Current folder**, **Current scheme**, **Wires selected in the scheme**, or **Choose schemes** as required. The effect of renumbering will be displayed only in the drawings based on your selection here.
- Select the **Remove wire numbers** radio button if you want to delete any specific types of wire numbers. Select the desired check box below it delete respective wire number.
- Select the **Number new wires and recalculate existing wire marks** radio button if you have performed any wire number formula change or added some new wires.
- Select the **Recalculate the wire numbers** radio button if you want to renumber all the wire numbers. If you select Recalculate manual check box under this radio button then only the manually assigned wire marks will be renumbered.
- Click on the **OK** button from the **Renumber wires** dialog box displayed. The wire numbering will be changed based on specified parameters; refer to Figure-7.

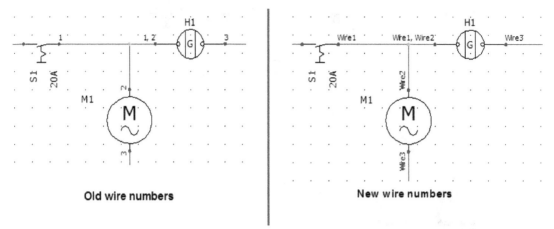

Figure-7. Schematic after changing wire number

HARNESSES

The **Harnesses** tool is used to create harness in electrical project which can later be used to create 3D electrical model. The procedure to create harness is given next.

- Click on the **Harnesses** tool from the **Management** panel in the **Project** tab of the **Ribbon**. The **Harness manager** will be displayed; refer to Figure-8.

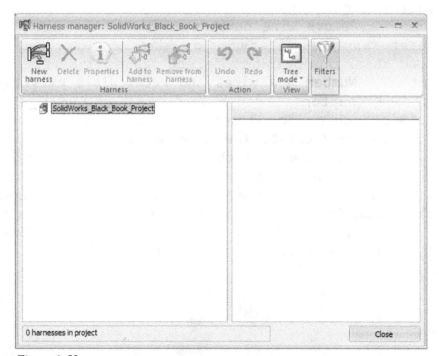

Figure-8. Harness manager

- Click on the **New harness** tool from the **Harness** panel in the **Harness manager**. The **Harness properties** dialog box will be displayed; refer to Figure-9.

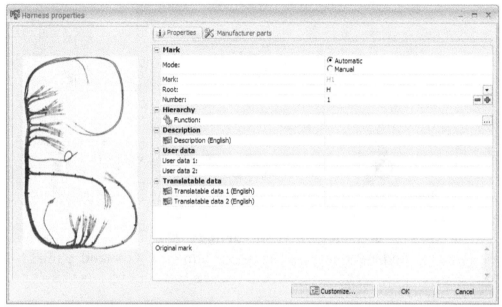

Figure-9. Harness properties dialog box

- Set the desired parameters in the **Properties** tab and select the desired manufacturer data from **Manufacturer parts** tab. Click on the **OK** button to create harness. A new harness will be created; refer to Figure-10.

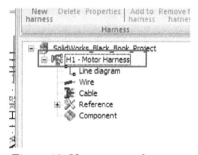

Figure-10. Harness created

- Close the **Harness manager** by clicking on the **Close** button.
- Select all the components and wires that you want to be added in harness from schematic or line diagram, and right-click on them. A shortcut menu will be displayed; refer to Figure-11.

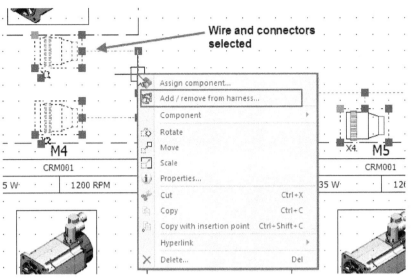

Figure-11. Shortcut menu for harness

- Click on the **Add/remove from harness** tool from the shortcut menu. The **Add/remove from harness** page of **Command** panel will be displayed with selected wires and components; refer to Figure-12.

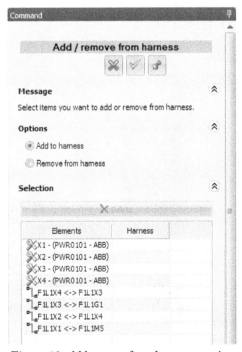

Figure-12. Add remove from harness page in Command panel

* Make sure the **Add to harness** radio button is selected and then click on the **OK** button from the panel. The Harness selector dialog box will be displayed asking you to select a harness.
* Select the desired harness from the dialog box and click on the **Select** button from the dialog box. The selected wires and components will be added to the harness.

If you check the harness in the **Harness manager** then you will find that components and wires are added to harness in respective categories; refer to Figure-13.

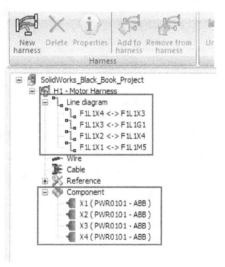

Figure-13. Wires and components added to harness

CABLE MANAGEMENT

You have learned about drawing cable in previous chapters. In this section, you will learn about managing cables. The procedure to do so is given next.

* Click on the **Cables** tool from the **Management** panel in the **Project** tab of the **Ribbon**. The **Cables manager** will be displayed; refer to Figure-14.

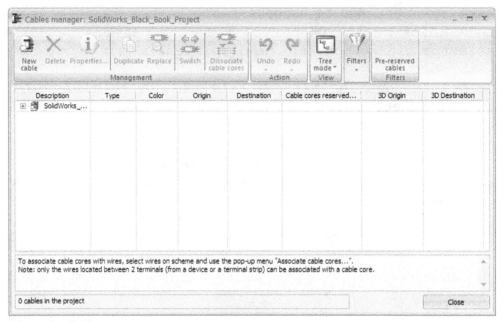

Figure-14. Cables manager

- Click on the **New cable** tool from the **Management** panel in the **Cables manager**. The **Cable references selection** dialog box will be displayed; refer to Figure-15.

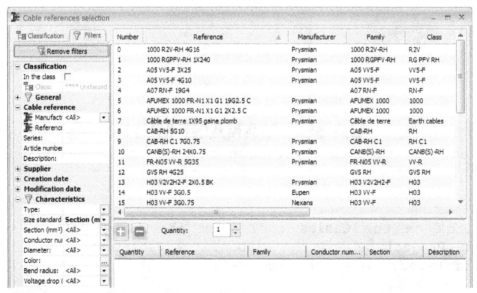

Figure-15. Cable references selection dialog box

- Set the desired number of conductors, size and other parameters in the **Filters** area of the dialog box. The available cables will be displayed in the right area of the dialog box.

- Double-click on the cable(s) that you want to use in your circuit and then click on the **Select** button from the dialog box. The cable(s) will be added in the **Cables manager**; refer to Figure-16.

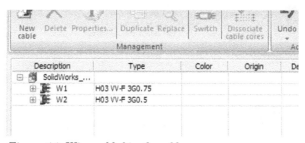

Figure-16. Wires added in the cable manager

- If you want to create more copies of a cable then select the cable from the **Cables manager** and click on the **Duplicate** button from the dialog box. Now, click on **Duplicate** button the number of times you want the selected cable.
- If you want to replace a cable with another cable then select the cable and click on the **Replace** button from the **Management** panel in the dialog box. The **Cable reference selection** dialog box will be displayed. Select the desired cable and click on the **Select** button from the dialog box.
- To set the length of cable and other parameters, select the cable in dialog box and click on the **Properties** button from the dialog box. The **Cable** dialog box will be displayed; refer to Figure-17. Set the length, color, voltage-drop, and other parameters of cable in the dialog box and click on the **OK** button. The properties will be applied to cable.
- Now, we have created cables but we have not mentioned what does it connect to. To connect the cables, we need wires draw in the schematic or block diagram. Close the **Cables manager** by clicking on the **Close** button. Select the wire(s) that you want to associate with a cable from the diagram and right-click on it. A shortcut menu will be displayed; refer to Figure-18.

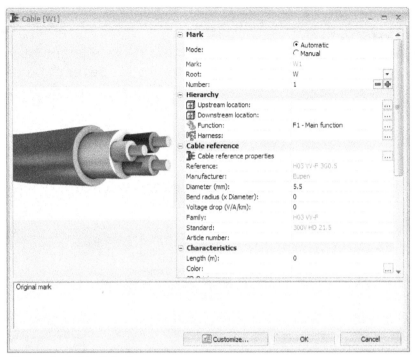

Figure-17. Cable dialog box

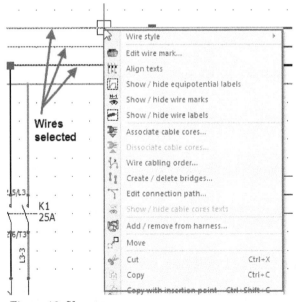

Figure-18. Shortcut menu

- Click on the **Associate cable cores** option from the shortcut menu. The **Associate cable cores** dialog box will be displayed; refer to Figure-19.

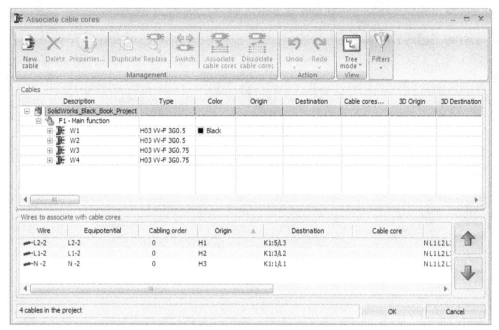

Figure-19. Associate cable cores dialog box1

- Expand the cable in dialog box to display its cores and select the cores to be associated. Click on the **Associate cable cores** button from the **Management** panel in the dialog box. The cable will be associated with selected wires; refer to Figure-20. Similarly, you can associate other cables. Click on the **OK** button to apply association.

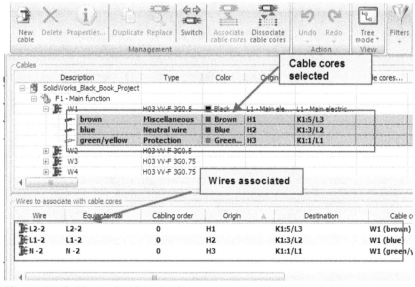

Figure-20. Cable associated with wires

WIRE CABLING ORDER

The **Wire Cabling Order** options are used to change the order by which the wires are connected to components. Take an example of the schematic given in Figure-21. In this figure, wire 1 coming from switch **S1** and wire 2 coming from motor **M** are getting joined at first terminal of glow bulb **H1** named at terminal as **1,2**. What if practically we need the two wires to join at motor terminal and not at glow bulb terminal. The procedure is given next.

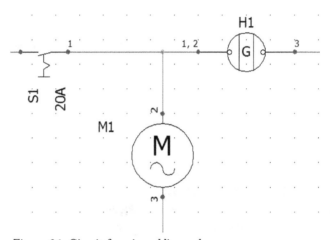

Figure-21. Circuit for wire cabling order

- By default, wire numbering is based on equipotential; refer to Figure-22. To change the wire numbering scheme for displaying individual wire numbers, click on **Wire styles** tool from the **Configuration** drop-down in the **Project** panel of the **Project** tab in the **Ribbon**. The **Wire style manager** dialog box will be displayed; refer to Figure-23.

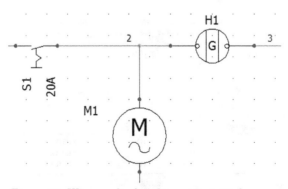

Figure-22. Wire number based on equipotential

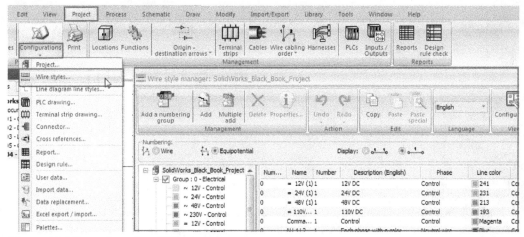

Figure-23. Wire styles tool and Wire style manager

- Select the **Wire** radio button from the **Numbering** area of the dialog box and click on the **Apply** button. The wire numbering scheme will get changed accordingly. Click on the **Close** button. Click on the **Renumber wires** tool and click **OK** from the dialog box if required.

- Click on the **Wire cabling order** tool from the **Wire cabling order** drop-down in the **Management** panel of the **Project** tab in the **Ribbon**; refer to Figure-24. The **Wire cabling order** dialog box will be displayed; refer to Figure-25.

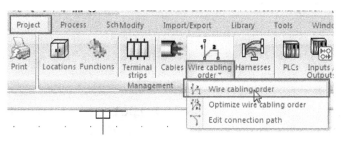

Figure-24. Wire cabling order tool

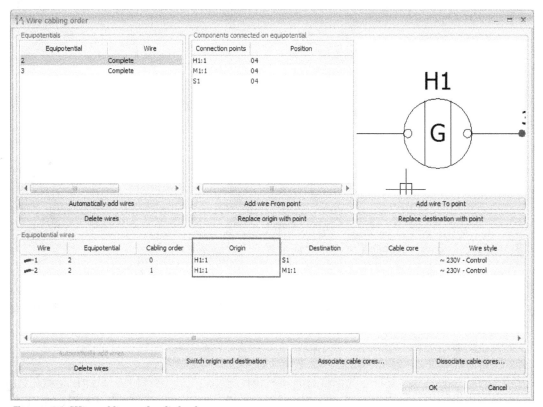

Figure-25. Wire cabling order dialog box

As you can see from above dialog box; the origin of both wires is terminal 1 of H1. But, we want the origin to be 1st terminal of M1. There are two operations required to achieve this: switch origin and destination of wire 2 and change the origin of wire 1. The steps to do so are given next.

- Select the wire **2** from the **Equipotential wires** area of the dialog box. The buttons at the bottom of dialog box will become active.
- Click on the **Switch origin and destination** button from the dialog box. Now, the origin of wire will become **M1:1**.
- Next, we want the origin of wire 1 as M1:1 to do so, click on the **M1:1** connection point from the **Components connected on eqiupotential** area of the dialog box drag-drop it on **H1:1** of wire **1** in **Equipotential wires** area of the dialog box; refer to Figure-26. The origin of wire will get changed; refer to Figure-27.

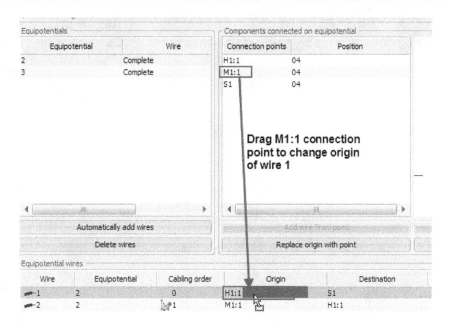

Figure-26. Dragging connection point

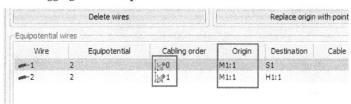

Figure-27. Wires with changed origin

Note that some icons are added in the **Cabling order** column in the **Equipotential wires** area of the dialog box. The icons indicate that the cabling order is changed manually. To set this wiring order as automatic, right-click on the icon and select the **Set as automatic cabling order** option from the shortcut menu. If you want to associate a cable to selected wire then select the wire and click on the **Associate cable cores** button from the dialog box. The **Associate cable cores** dialog box will be displayed. Rest of the procedure is same as discussed earlier.

PLC MANAGER

PLC Manager is used to manage all PLCs in the project at one place. The procedure to use this tool is given next.

- Click on the **PLCs** tool from the **Management** panel of the **Project** tab in the **Ribbon**. The **PLCs manager** will be displayed; refer to Figure-28.

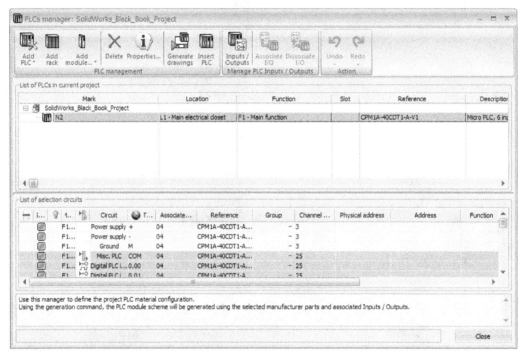

Figure-28. PLCs manager

- Click on the **Insert PLC** tool if you have not inserted a PLC earlier. The procedure is same as discussed earlier in the book.

Adding Racks

- If you want to add a new rack to the PLC then select the PLC from the **PLCs manager** and click on the **Add rack** button from the **Management** panel in the dialog box. The **Manufacturer part selection** dialog box will be displayed. Select the desired manufacturer part using the dialog box and click on the **Select** button. The rack will be added to the selected PLC system; refer to Figure-29.

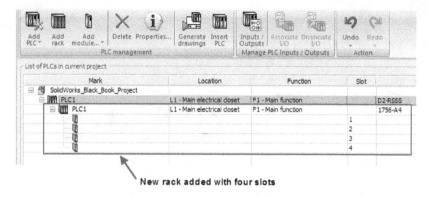

New rack added with four slots

Figure-29. Rack added to PLC

Adding PLC Modules

- If you want to add plc module without interface to the selected rack slot then click on the **Add module** tool from the **Add module** drop-down. If you want to add a plc module with interface then click on the **Add module with interface** tool from the **Add module** drop-down in the **PLC management** panel of the **PLCs manager**. The **Manufacturer part selection** dialog box will be displayed.
- Select the desired module using the filters and click on the **Select** button from dialog box. The PLC module will be added to the rack.

Generating PLC Drawings

- If you want to generate drawings for PLCs in **PLCs manager** then select the desired PLC(s) and click on the **Generate drawings** tool from the **PLC management** panel in **PLCs manager**. The **Selection of: Books, Folders** dialog box will be displayed; refer to Figure-30. Select the desired book of the project and click on the **OK** button. The PLC drawings will be added in the project book and **Report creation** box will be displayed. Click on the **Close** button from the **Report creation** box to exit.

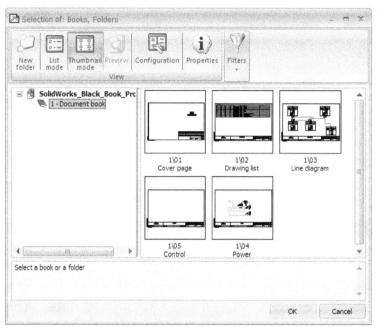

Figure-30. Selection of: Books, Folders dialog box

Managing Inputs/Outputs of PLCs

• Click on the **Inputs/Outputs** tool from the **Manage PLC Inputs/ Outputs** panel of the **PLCs manager**. The **Inputs/Outputs manager** will be displayed; refer to Figure-31.

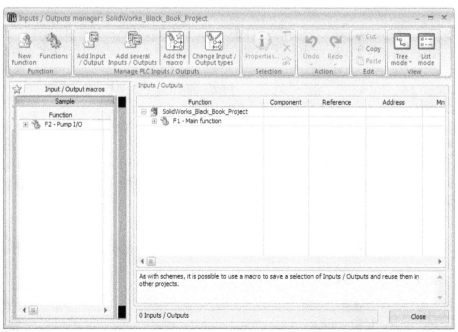

Figure-31. Inputs/Outputs manager

Adding Function

- We divide various inputs & outputs of PLCs based on their functions. So, first we need to create functions for input-output. Select the function or book in which you want to add the function and click on the **New function** tool from the **Function** panel in the dialog box. The **Function** dialog box will be displayed; refer to Figure-32.

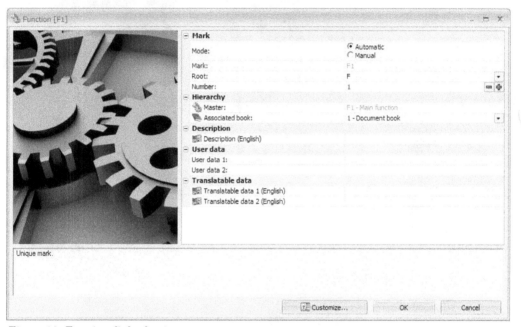

Figure-32. Function dialog box1

- Select the **Manual** radio button from **Mark** node and specify the desired mark like Motor, Conveyor, and so on in the **Mark** edit box. Click on the **OK** button from the dialog box. The function will be added; refer to Figure-33.

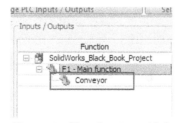

Figure-33. New function added

Adding Input/Output

- Select the function to which you want to add input/output circuits and click on the **Add Input/Output** button from **Manage PLC Inputs/Outputs** panel of the dialog box. A drop-down with input-output options will be displayed; refer to Figure-34.

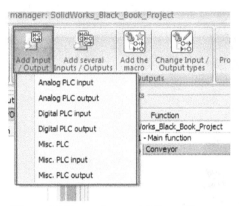

Figure-34. Drop-down with Input/Output options

- Select the desired option from the drop-down. The respective input/output will be added to the selected function; refer to Figure-35.

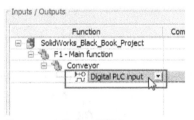

Figure-35. Digital PLC input added

- If you want to add multiple inputs/outputs then click on **Add several Inputs/Outputs** button from the **Manage PLC Inputs/ Outputs** panel of the dialog box. A drop-down similar to the one shown for **Add Input/Output** option will be displayed.
- Select the desired option from drop-down (We have selected the **Digital PLC input** option). The **Multiple insertion** dialog box will be displayed; refer to Figure-36.

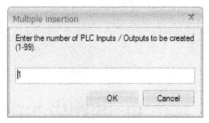

Figure-36. Multiple insertion dialog box

- Specify the number of inputs/outputs to be added in the edit box and click on the **OK** button. The specified number of inputs/outputs will be added. Similarly, add the other inputs/outputs to PLC.
- If you want to edit properties of any of the input or output then select it and click on the **Properties** button from the **Selection** panel in the dialog box. The **Input/ Output properties** dialog box will be displayed; refer to Figure-37. Specify the desired properties and click on the **OK** button.

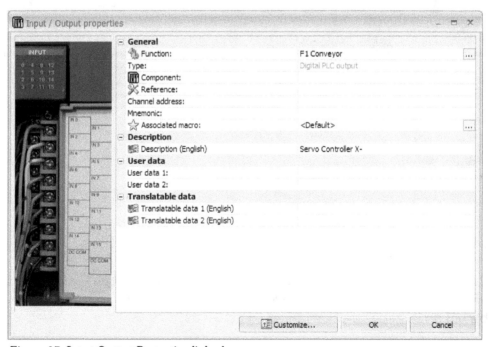

Figure-37. Input Output Properties dialog box

- Close the dialog box once, you have specified desired number of inputs and outputs.

Till now we have created PLCs and we have created functions with inputs and outputs, but we have not specified which terminal of PLC does which work. So, we will now connect PLC terminals with Inputs/Outputs specified for PLC.

Connecting PLC Terminals with Input/Output Circuits

- Click again on the **PLCs** tool from the **Management** panel in the **Project** tab of the **Ribbon** to display **PLCs manager**. Select the PLC whose terminals are to be connected. The terminals will be displayed in the bottom of the PLCs manager; refer to Figure-38.

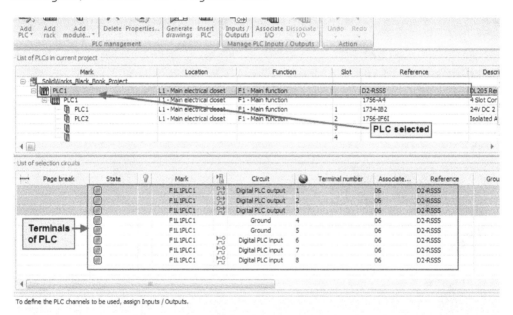

Figure-38. PLC terminals in PLCs manager

- Select the terminals with Digital PLC output circuits and right-click on them. A shortcut menu will be displayed; refer to Figure-39.
- Click on the **Assign an existing PLC Input/Output** option from shortcut menu. The **Inputs/Outputs selection** dialog box will be displayed; refer to Figure-40.

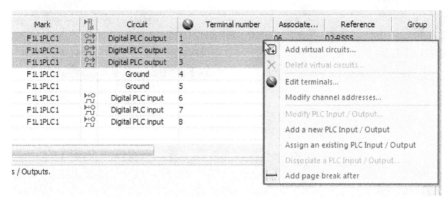

Figure-39. Shortcut menu for PLC terminals

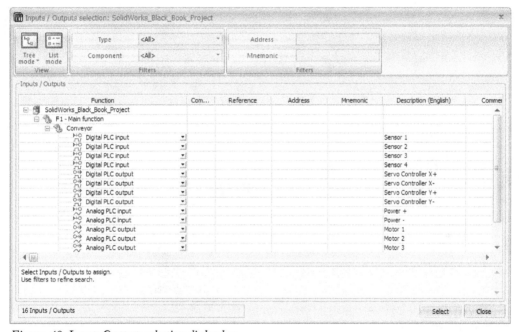

Figure-40. Inputs Outputs selection dialog box

- Since, we have selected 3 terminals with Digital PLC output circuits in Figure-39 so we need to select 3 Digital PLC outputs from the dialog box. Select the Outputs from the dialog box while holding the **CTRL** key and click on the **Select** button. An information box will be displayed confirming the association. Click on the **OK** button.

Similarly, set the other inputs and outputs of PLC terminals. Now, click on the **Generate drawings** tool. If you already inserted PLC drawings then the **Dynamic PLC insertion** dialog box will be displayed; refer to Figure-41.

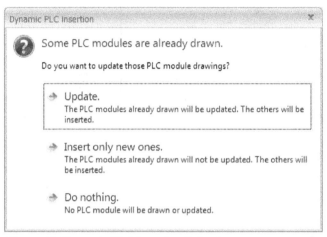

Figure-41. Dynamic PLC insertion dialog box

Click on the **Update** button. Select the book for inserting new drawings and click on the **OK** button from the **Selection of: Books, Folders** dialog box. Close the **Report creation** dialog box and other dialog boxes. The PLC drawings will be added and updated.

TITLE BLOCKS MANAGER

The **Title blocks manager** tool is used to create and manage title blocks. Using this tool you can create title blocks as per your requirement. The procedure to create title blocks is given next.

• Click on the **Title blocks manager** tool from the **Graphics** panel in the **Library** tab of the **Ribbon**. The Title blocks manager dialog box will be displayed; refer to Figure-42.

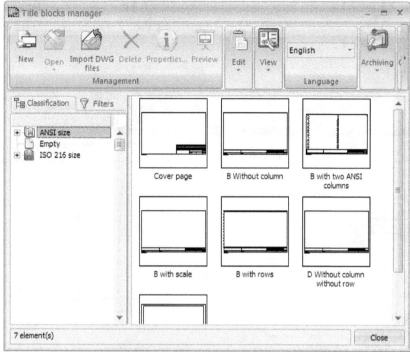

Figure-42. Title blocks manager dialog box

- To edit any title block template, double-click on it from the **Title blocks manager.** The options for editing block will be displayed; refer to Figure-43.

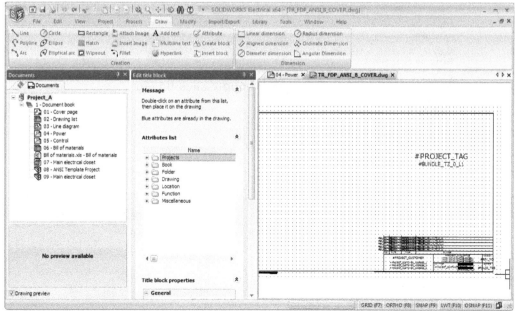

Figure-43. Options for editing block

- Using the sketching tools available in the **Draw** tab of **Ribbon,** you can create the boundary of the template; refer to Figure-44.

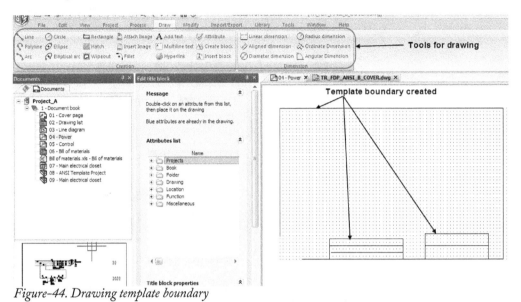

Figure-44. Drawing template boundary

- Expand the categories in Attribute list and double click on the desired attribute tag to insert it in the title block; refer to Figure-45.

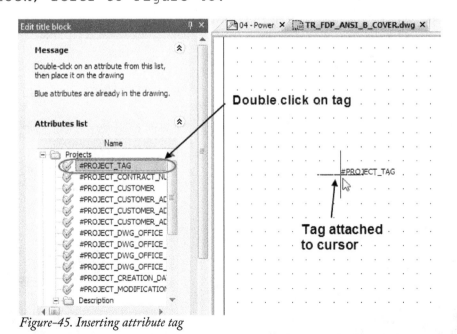

Figure-45. Inserting attribute tag

- Click at the desired location in the drawing to place the tag.
- Expand the **Title block properties** rollout and specify the properties of title block in the fields; refer to Figure-46.

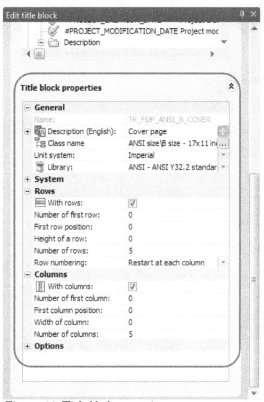

Figure-46. Title block properties

- Save the template drawing file and close it.

SYMBOLS MANAGER

The **Symbols manager** tool as the name suggests is used to manage symbols of electrical database. Using this tool, you can edit an earlier created symbol or you can create a new symbols as required. The procedure to use this tool is given next.

- Click on the **Symbols manager** tool from the **Graphics** panel in the **Library** tab of the **Ribbon**; refer to Figure-47. The **Symbols manager** dialog box will be displayed as shown in Figure-48.

Figure-47. Symbols manager tool

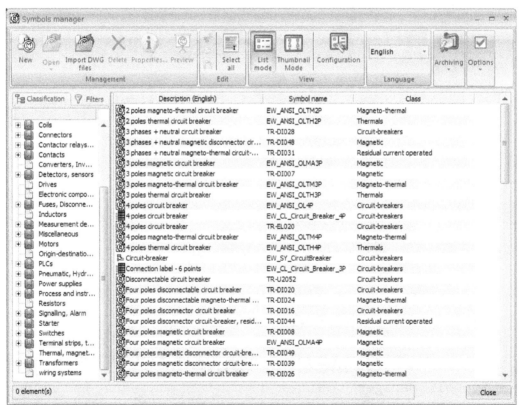

Figure-48. Symbols manager dialog box

- To edit any symbol, double-click on it from the dialog box. Editing environment will be displayed along with the symbol attributes; refer to Figure-49.

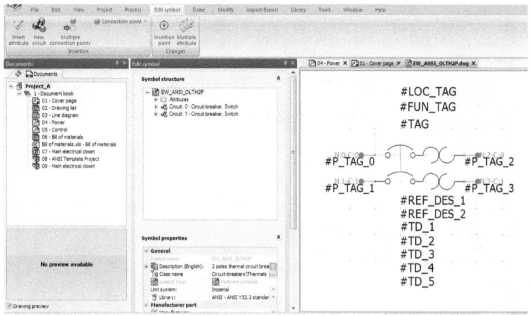

Figure-49. Editing environment for symbol

- Using the tools in the **Edit symbol** tab in **Ribbon,** you can add connection points, new circuits, attributes and so on. We will learn about the tools in Edit symbol tab later.
- Change the attributes and properties of symbol as done for title block.
- Save the symbol drawing file and close it.

Creating New Symbol

- To create a new symbol, click on the **New** button from the **Management** panel of **Symbols manager** dialog box; refer to Figure-50.
- Click in the **Symbol name** field and specify the desired name for the symbol.
- Click on the Ellipse button for **Class name** field. The **Class selector** dialog box will be displayed; refer to Figure-51.

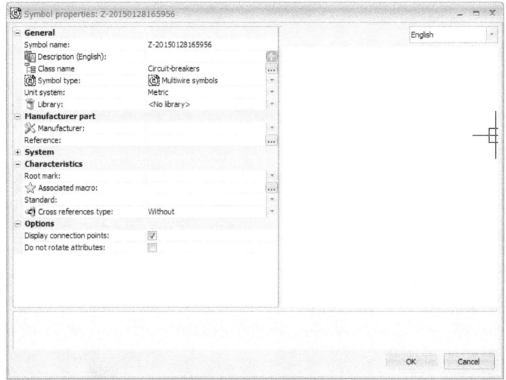

Figure-50. Symbol properties dialog box

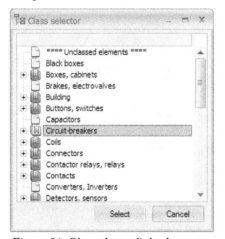

Figure-51. Class selector dialog box

- Select the desired category for the symbol and click on the **Select** button.
- Click in the **Symbol type** drop-down and select the type of symbol; refer to Figure-52.

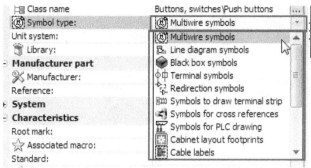

Figure-52. Symbol type drop-down

- Similarly, specify other properties of the symbol and click on the **OK** button from the dialog box. The symbol will be added in the library; refer to Figure-53.

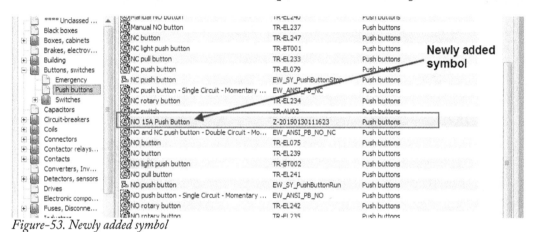

Figure-53. Newly added symbol

- Double-click on the newly added symbol. The symbol editing environment will be displayed with blank drawing area.
- Draw the symbol by using the tools available in the **Draw** tab of **Ribbon**; refer to Figure-54.

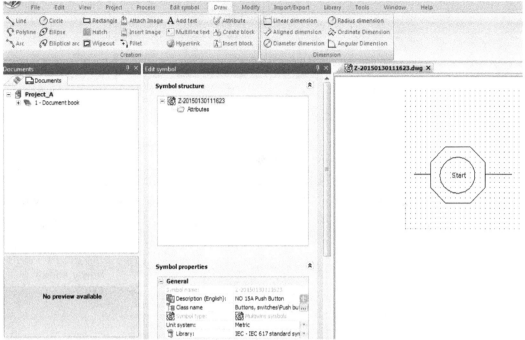

Figure-54. Drawing new symbol

- Click on the **New circuit** tool from the **Insertion** panel in the **Edit symbol** tab of **Ribbon**. The **New circuit** dialog box will be displayed; refer to Figure-55.

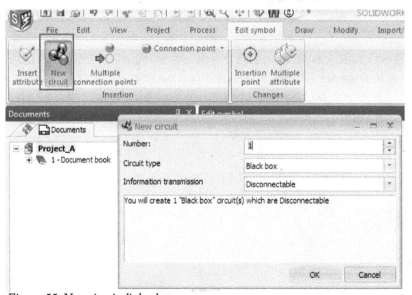

Figure-55. New circuit dialog box

- Specify the number of circuits you want to create for the current symbol by using spinner in the dialog box.
- From the **Circuit type** drop-down, select the circuit (For our case its NO button) and set the transmission information.
- After specifying the settings, click on the **OK** button from the dialog box. The circuit will be added to the symbol structure.
- Click on the down arrow next to **Connection point** button in the **Insertion** panel. A list of tools will be displayed; refer to Figure-56.

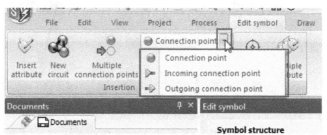

Figure-56. Tools for creating connection points

- The **Incoming connection point** tool is used to set inlet for the component and similarly, **Outgoing connection point** tool is used to set outlet for the component. If the component can be connected by any orientation then the **Connection point tool** is used. Click on the desired tool from the drop-down (**Connection point** tool is selected in our case). You are asked to select the reference point on the symbol created.
- Click to specify the connection point. Make sure that you have selected the **OSNAP** button to activate Object snapping for easy point selection; refer to Figure-57.

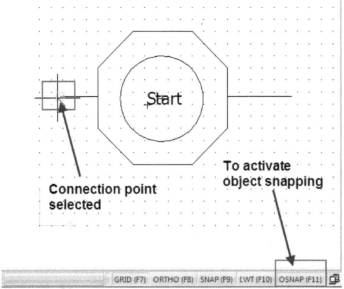

Figure-57. Object snapping

- Similarly, set the connection point on the other side of the symbol. Note that the tag are also attached to the connection points automatically.
- Click on the **Insert attribute** button to insert attributes of symbol. The **Attribute management** dialog box will be displayed; refer to Figure-58.

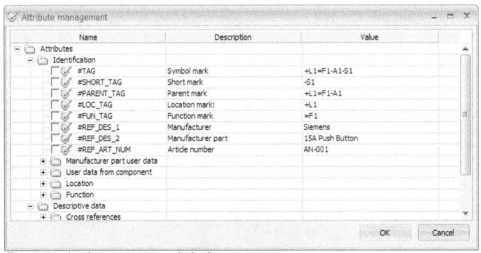

Figure-58. Attribute management dialog box

- To add an attribute to the component, select the respective check box from the dialog box and click on the **OK** button from the dialog box. The tag will get attached to the cursor.
- Click at the desired position near the symbol to place the tag.
- Now, click on the **Insertion point** tool from the **Changes** panel in the **Edit symbol** tab to specify the insertion point for the symbol. You are asked to select a point.
- Click at the desired location on the symbol to specify it as insertion point. Generally, the best location is one of the connection points earlier specified.
- Now, save the drawing file of symbols and close it. You are ready to use this symbol in drawings.

2D FOOTPRINTS MANAGER

2D footprints are used in designing the used control panel. The 2D footprints manager is used to create and manage footprints. The procedure to use the 2D footprints manager is discussed next.

- Click on the **2D footprints manager** tool from the **Graphics** panel in the **Library** tab of **Ribbon**. The **Cabinet layout footprints manager** dialog box will be displayed; refer to Figure-59. Here, we will discuss the procedure to create new footprint. You can use the same parameters to edit the footprint.

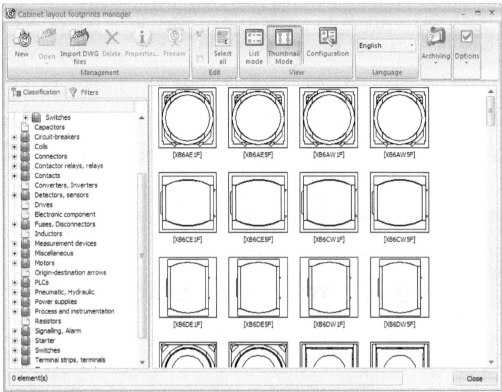

Figure-59. Cabinet layout footprints manager dialog box

• Select the category from the **Classification** pane in the left of dialog box to which our new footprint belongs to. Like, we have selected Electronic component category in our case.

In the same way, you can use the **Cable references manager** to create cable references and **Manufacturer parts manager** to manage the data related to manufacturer parts.

Chapter 6

Cabinet Layout

Topics Covered

The major topics covered in this chapter are:

- *Introduction*
- *2D Cabinet Layout*
- *Inserting 2D Footprint*
- *Inserting Terminals Strips and Rails*
- *Inserting Ducts*
- *Adding cabinets*
- *Aligning footprints*

INTRODUCTION

In the previous chapters, you have learnt to create schematic circuit diagrams. After creating those circuit diagrams, the next step is to create panels. A panel is the box consisting of various electrical switches and PLCs to control the working of equipment. Refer to Figure-1. Note that the panel shown in the figure is back side panel of a machine. This panel is generally hidden from the operator. What an operator see is different type of panel; refer to Figure-2. We call this panel as User Control panel. In both the cases, the approach of designing is almost same but the interaction with the user is different. The User Control Panel is meant for Users so it can have push buttons, screen, sensors, key board and so on. On the other side, the back panel will be having relays, circuit breakers, sensors, connectors, plcs, switches and so on.

Figure-1. Panel

Figure-2. User Control panel

If we start linking the schematic drawings with the panel drawings then the common platform is the component tag and the location code. Suppose we have created a push button in the schematic with tag -04PB2 then in the panel layout you should insert the same push button with the same tag. The location of the Push button will be decided by the Location code. The components that are having same location code should be placed at the same place in the panel. Also, the components that are having the same Function code should be placed together in the cabinet.

There are two options in SolidWorks Electrical to create cabinet layout; **2D cabinet layout** and **SOLIDWORKS cabinet layout**. The **2D cabinet layout** tool is used to create 2 dimensional layout of the cabinet. The **SOLIDWORKS cabinet layout** tool is used to create 3D model of the cabinet. The procedures to use these tools are discussed next.

2D CABINET LAYOUT

The **2D cabinet layout** tool is available in the **Processes** panel of the **Process** tab in the **Ribbon**; refer to Figure-3. The procedure to use this tool is given next.

Figure-3. 2D cabinet layout tool

- Click on the **2D cabinet layout** tool from the **Processes** panel. The **Create 2D cabinet layout drawings** dialog box will be displayed; refer to Figure-4.

Figure-4. Create 2D cabinet layout drawings dialog box

- Select the check box for only those locations for which you want to make the cabinets and then click on the **OK** button. New cabinet drawing/drawings will be added in the **Documents browser**; refer to Figure-5.

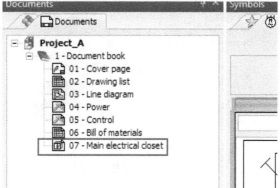

Figure-5. Cabinet layout drawing added

- Double-click on the newly created cabinet drawing. The Cabinet layout editing environment will be displayed; refer to Figure-6.

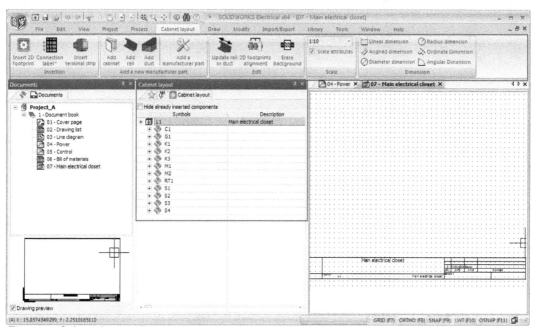

Figure-6. Cabinet layout editing environment

- To insert footprints of components already created in Line diagram and Schematic diagram, expand the node of respective component from the **Cabinet layout browser** and select the corresponding check box; refer to Figure-7. The **2D footprint insertion** page of **Command** panel will be displayed and the footprint will get attached to the cursor; refer to Figure-8.

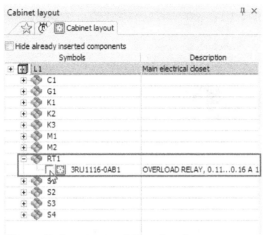

Figure-7. Auto generated footprints for components

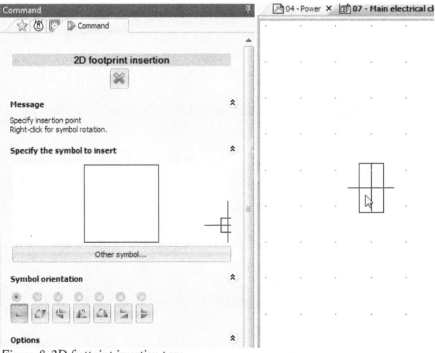

Figure-8. 2D footprint insertion page

- If you want to use another footprint symbol then click on the **Other symbol** button from the **Command** panel and select the desired symbol.
- Click in the drawing to place the footprint. Similarly, you can place other footprints.

The tools available in the **Cabinet layout** tab are discussed next.

Insert 2D footprint

This tool is used to insert footprints for the components existing the Line diagram and Schematic drawing of the project. The procedure to use this tool is given next.

- Select a component from the **Cabinet layout browser**.
- Click on the **Insert 2D footprint** tool from the **Insertion** panel in the **Cabinet layout** tab of the **Ribbon**. Related footprint will get attached to cursor and the **2D footprint insertion** page will be displayed in the **Command** panel.
- Click in the drawing area to place the footprint.

The tools available for Connection labels in **Cabinet layout** tab work in the same way as discussed earlier.

Insert terminal strip

The terminals are used to connect components with supplies. Use of terminals ensure the safety of wires since in case of short-circuit, wire will get burnt at the terminal. The terminal strip comprises of many terminals. The procedure to insert terminal strips is given next.

- Click on the **Insert terminal strip** tool from the **Insertion** panel in the **Cabinet layout** tab of the **Ribbon**. The **Terminal strip selector** dialog box will be displayed; refer to Figure-9.
- Click on the **New** tool from the **Management** panel in the dialog box. The **Component properties** dialog box will be displayed; refer to Figure-10.

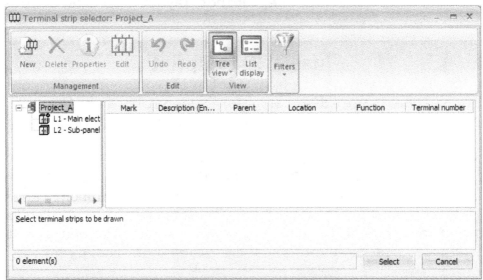

Figure-9. Terminal strip selector dialog box

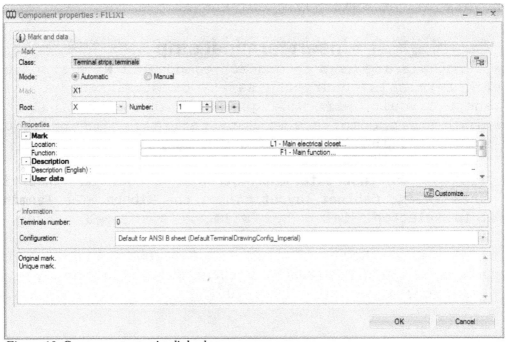

Figure-10. Component properties dialog box

- Specify desires description and properties for the terminal strip and click on the **OK** button from the dialog box. The terminal strip will get added in the **Terminal strip selector** dialog box.

- Select the newly created strip from the dialog box and click on the **Edit** button. The **Terminal strip editor** dialog box will be displayed as shown in Figure-11.

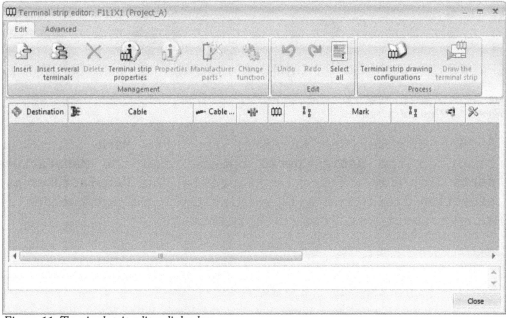

Figure-11. Terminal strip editor dialog box

- Click on the **Insert** tool from the **Management** panel in the **Edit** tab of the dialog box. A terminal will get added in the terminal strip.
- To add multiple strip at one time click on the **Insert several terminals** tool from the **Management** panel. The **Multiple insertion** dialog box will be displayed as shown in Figure-12.

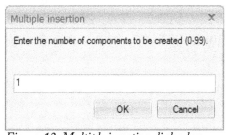

Figure-12. Multiple insertion dialog box

- Specify the desired number of terminals to be added in the strip and click on the **OK** button. Like in Figure-13 we have created 6 terminals in a terminal strip.

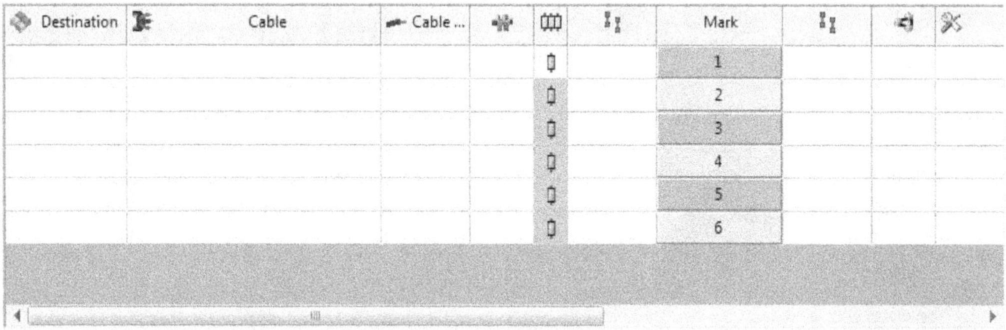

Destination		Cable	Cable ...				Mark			
							1			
							2			
							3			
							4			
							5			
							6			

Figure-13. Terminal strip created

- Select the desired terminal from the **Mark** column and click on the **Assign parts** button from the **Manufacturer parts** drop-down; refer to Figure-14. The **Manufacturer part selection** dialog box will be displayed. Select the desired manufacturer part for each terminal.

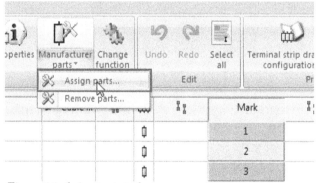

Figure-14. Assign parts tool

- To create a drawing file of terminal strip with specified configuration, click on the **Terminal strip drawing configurations** tool from the **Process** panel. The **Terminal strip drawing configuration** dialog box will be displayed; refer to Figure-15.

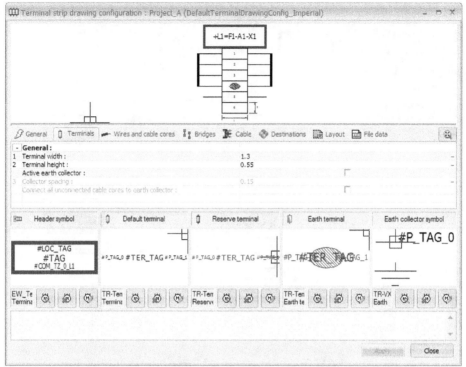

Figure-15. Terminal strip drawing configuration dialog box

- Set the parameters for the terminal strip and click on the click on the **Apply** button.
- Click on the **Close** button to exit the dialog box.
- If you want to create a separate drawing for the terminals then click on **Draw the terminal strip** tool. The **Selection of** dialog box will be displayed; refer to Figure-16.
- Select the desired book and click on the **OK** button from the dialog box. The drawing will be added in the selected book and summary will be displayed.
- Close the summary box and click on the **Close** button from the **Terminal strip editor** dialog box.

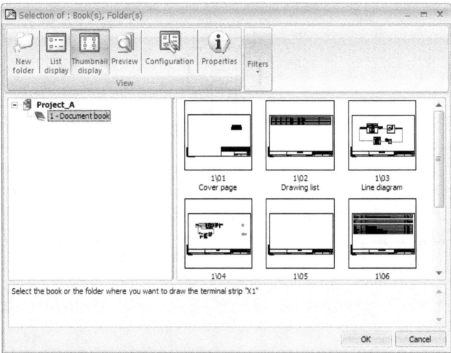

Figure-16. Selection of dialog box

- Now, click on the **Select** button from the **Terminal strip selector** dialog box. A terminal will get attached to the cursor; refer to Figure-17.

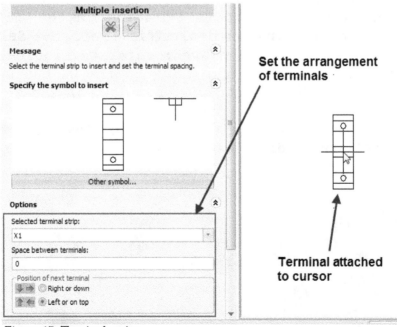

Figure-17. Terminal setting

- Set the desired parameters for successive terminals in the **Options** rollout in **Command** panel displayed.
- Click to specify the first terminal, rest of the terminals will be created automatically; refer to Figure-18.

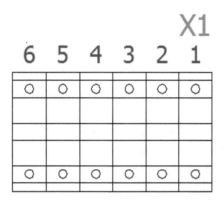

Figure-18. Terminals created

Add cabinet

Cabinet is an enclosure used to pack all the panel components so that the components inside cabinet are safe from external factors like dirt, water and so on. The procedure to add a cabinet is given next.

- Click on the **Add cabinet** tool from the **Add a new manufacturer part** panel in the **Cabinet layout** tab of the **Ribbon**. The **Manufacturer part selection** dialog box will be displayed as discussed earlier.
- Set the desired filters in the dialog box and click on the **Search** button to display the available cabinets; refer to Figure-19.

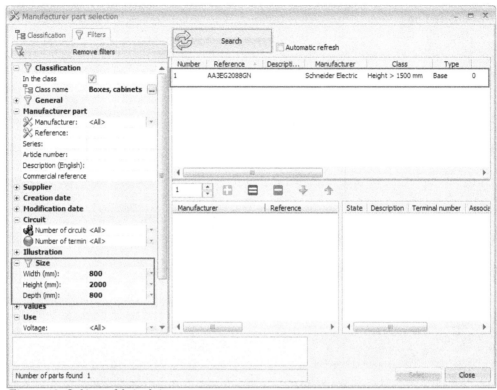

Figure-19. Cabinet of desired size

- Double-click on the cabinet to be used and then click on the **Select** button. The cabinet will get attached to the cursor.
- Click to place the cabinet at the desired location.

Although we have discussed this tool later, but this is the first step when we start working on cabinet layout. So, while practicing in your organization, you should first place the cabinet, then rails and then the other components. Now, we will discuss about inserting rails in the cabinet.

Inserting rail

Rail is a strip of metal which holds various components like circuit breakers, contacts and so on inside the cabinet. The procedure to insert rails in cabinet is given next.

- Click on the **Insert rail** tool from the **Add a new manufacturer part** panel in the **Ribbon**. The **Manufacturer part selection** dialog box will be displayed.
- Set the desired filter and click on the **Search** button. The list of rails available in the database will be displayed.
- Double-click on the desired rail and then click on the **Select** button from the dialog box. The rail strip will get attached to the cursor.
- Specify the starting point of the rail by clicking inside the cabinet. You are asked to specify the length of the rail; refer to Figure-20.

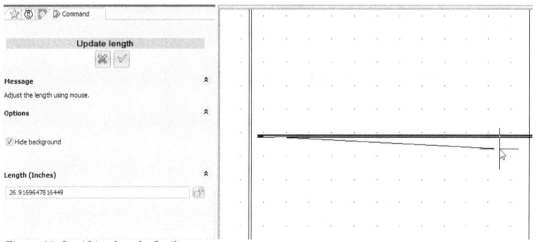

Figure-20. Specifying length of rail

- Click to specify the end point of the rail or enter the desired length in the **Command** panel. The rail will be created.
- Repeat the steps to create multiple rails or you can Copy the rail by selecting it and pressing CTRL+C from keyboard and CTRL+V to paste the rail. Click to place the copied rail.

Inserting Ducts

Ducts are inserted in the same way as rails are inserted. To insert the ducts, click on the **Add duct** tool from the **Add a new manufacturer part** panel. Rest of the procedure is similar to inserting rails.

Aligning Footprints

Once you have created footprints of components in the cabinet, the next step is to align them properly. In SolidWorks Electrical there is a separate tool to perform this task. The procedure to align footprints is given next.

- Select the footprints that you want to align either by window selection or by holding the CTRL key while selecting footprints.
- Click on the **2D footprints alignment** tool from the **Edit** panel in the **Cabinet layout** tab of the **Ribbon**. The **Align parts** page will be displayed in the **Command** panel; refer to Figure-21.

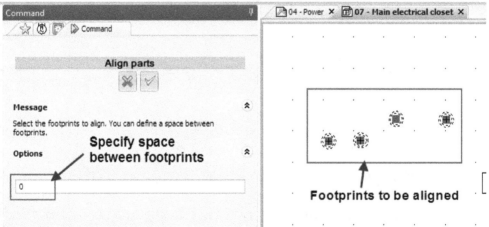

Figure-21. Aligning footprints

- Specify the spacing between consecutive footprints in the edit box available in the **Command** panel and click on the **OK** button. The footprints will get aligned by specified distance; refer to Figure-22.

Figure-22. Aligned footprints

Note that once you are done with the designing of cabinet, its important to apply the dimensions so that it can be manufactured.

Till this point, we have learnt almost all the tools necessary for the use of SolidWorks Electrical in industry. In the next chapter, we will perform some exercises using SolidWorks Electrical.

FOR STUDENT NOTES

Chapter 7

Practice and Practical

Topics Covered

The major topics covered in this chapter are:

- ***Practice questions and practical***

INTRODUCTION

In the previous chapters, we have learnt so many tools but it is easy to follow the procedure give in the book and perform the do this and do that format. Does it really help us in our general work in industry? My answer would be No. In this chapter, we will practically apply the tools and techniques to understand the application of SolidWorks Electrical in the industry.

PRACTICAL 1

Starting with the most basic control circuit, we will create a 3-Wire Control circuit for low voltage protection as given in Figure-1.

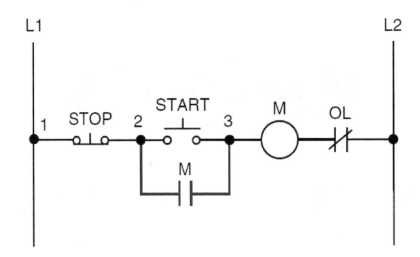

Figure-1. Practical 1

Steps:

Starting a new Project

- I hope you have started SolidWorks Electrical till now!! Click on the **Projects manager** tool from the **SolidWorks Electrical** panel in the **File** tab of the **Ribbon** to display the **Projects Manager**; refer to Figure-2.

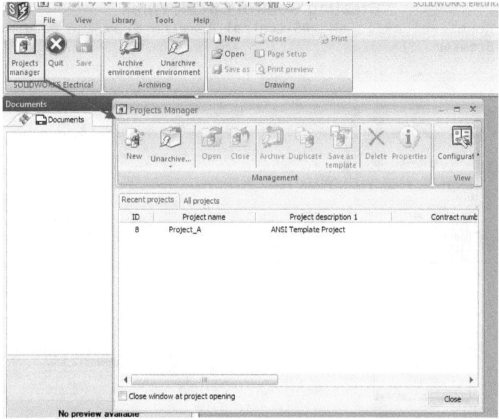

Figure-2. Projects Manager

- Click on the **New** button from the **Management** panel in the **Projects Manager.** You will be asked to select the desired template from the **Create a new project** dialog box.
- Select the ANSI template and click on the OK button from the dialog box. The **Project language** dialog box will be displayed.
- Select **English** as language and click on the **OK** button from the dialog box. The **Project** dialog box will be displayed.
- Specify the user data as per the instructions given by your tutor.
- Once you click on the **OK** button after specifying the data, the newly created project gets added in the project list of **Projects Manager.**
- Make sure that the newly created project is open and then close the **Projects Manager** by selecting the **Close** button.

Creating Control Circuit Wires

- Double-click on the **05-Control** drawing from the **Documents Browser.** The drawing will open; refer to Figure-3.

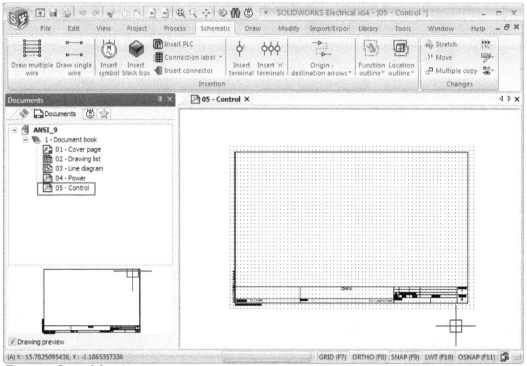

Figure-3. Control drawing

- Click on the **Draw single wire** tool from the **Insertion** panel in the **Schematic** tab of the **Ribbon**. The **Electrical wires** page will be displayed in the **Command** panel.
- Enter the number of lines as **2** in the **Number of lines** spinner cum edit box in the **Command** panel; refer to Figure-4.

Figure-4. Number of lines setting

- Click in the **Space between lines** edit box and specify the value as **8**.
- Click in the drawing area and create the wire lines as shown in Figure-5 and exit the tool.

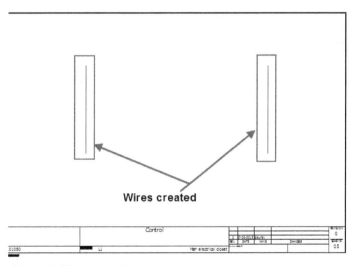

Figure-5. Wires created

- Click again on the **Draw single wire** tool and set the number of lines as **1**.
- Draw a wire connecting the earlier created wires as shown in Figure-6.

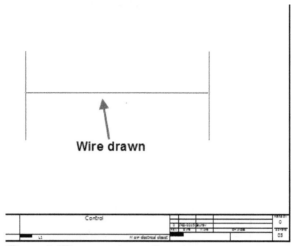

Figure-6. Wire connecting earlier wires

- Similarly, create a branch circuit in the connecting wire as shown in Figure-7.

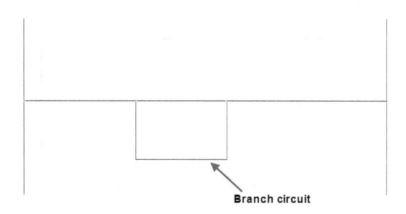

Figure-7. Creating branch circuit

Placing components

- Click on the **Insert symbol** tool from the **Insertion** panel in the **Schematic** tab of the **Ribbon**. The **Symbol insertion** page will be displayed in the **Command** panel.
- Click on the **Other symbol** button from the page. The **Symbol selector** dialog box will be displayed; refer to Figure-8.

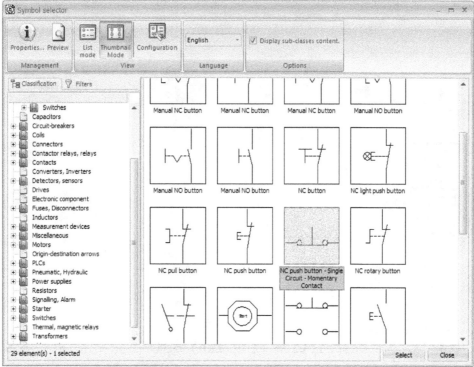

Figure-8. Symbol selector dialog box

- Select the **NC push button-Single Circuit - Momentary Contact** symbol from the **Buttons, switches > Push buttons** category in the dialog box and click on the **Select** button. The symbol will get attached to the cursor.
- Click on the wire to place the symbol at location as shown in Figure-9. The **Symbol properties** dialog box will be displayed as shown in Figure-10.

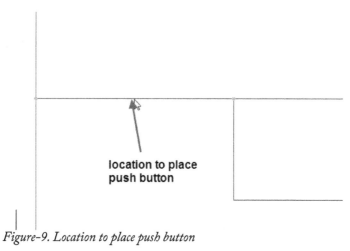

Figure-9. Location to place push button

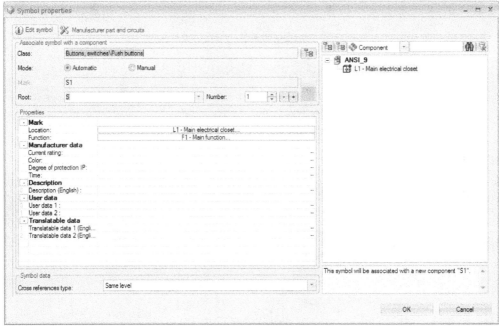

Figure-10. Symbol properties dialog box

- Click in the **Description (English)** edit box and enter **STOP**.
- Note that L1 is the location code for the symbol which means that while manufacturing, the component will be placed in L1 - Main electrical closet location.
- Click on the **OK** button from the dialog box. The symbol will be placed.
- Again, click on the **Insert symbol** button from the **Insertion** panel in the **Ribbon** and place the **NO push button - Single Circuit - Momentary Contact** symbol as shown in Figure-11. Specify the description as **START**.

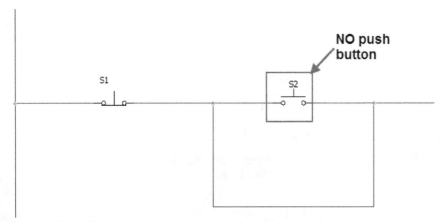

Figure-11. NO Push button placed

- Similarly, place the other symbols in the circuit; refer to Figure-12.

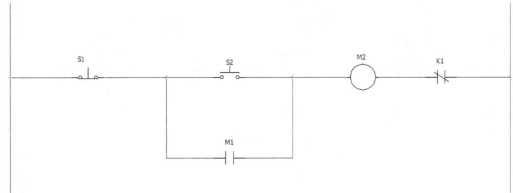

Figure-12. Drawing after placing all symbols

PRACTICE SET 1

On the basis of above practical, below are some drawings given in Figure-13, Figure-14, Figure-15, and Figure-16 that you can practice:

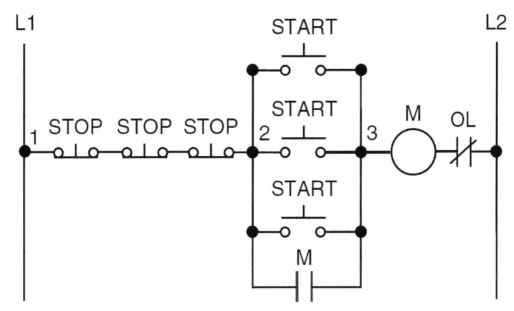

Figure-13. Practice 1

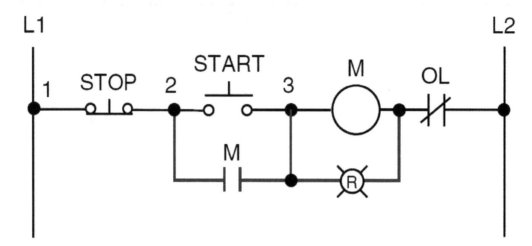

Figure-14. Practice 2

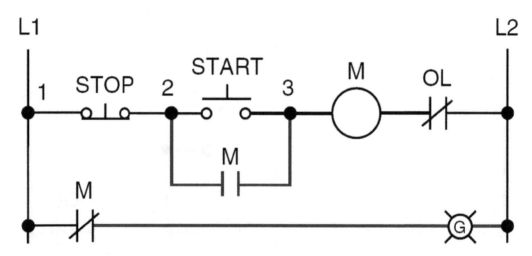

Figure-15. Practice 3

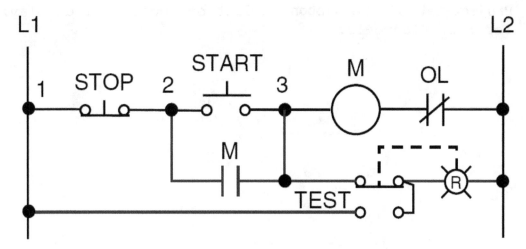

Figure-16. Practice 4

PRACTICAL 2

Create a transformer circuit with fuses as shown in Figure-17.

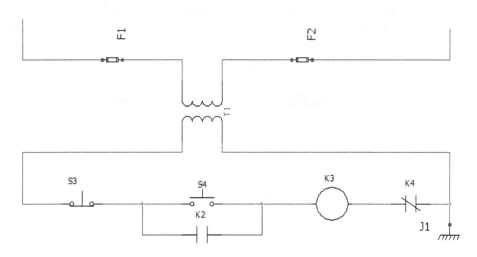

Figure-17. Practical 2

Starting New Drawing

I hope you are working in the same project that we have started in the Practical 1. Now, we will add another drawing the project.

- Open the project created in Practical 1. Click on the down arrow below **New** button in the **Project** panel in the

Project tab of the **Ribbon**. A list of tools will display; refer to Figure-18.

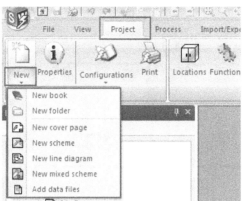

Figure-18. Tools for creating new drawings

- Click on the **New scheme** tool from the list. The Drawing dialog box will be displayed.
- Click in the **Description (English)** edit box and specify the description as **Transformer fused circuit**.
- Click on the **OK** button from the dialog box. A new drawing will be added in the **Documents browser**. Double-click on the newly created drawing to open it; refer to Figure-19.

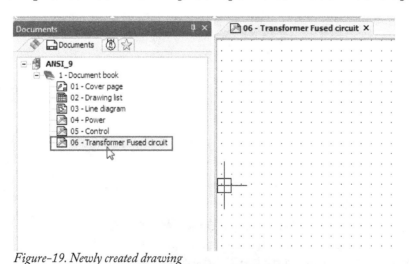

Figure-19. Newly created drawing

Inserting Wires

- Click on the **Draw single wire** tool from the Insertion panel in the Schematic tab of the Ribbon and draw a wire 230V AC wire as shown in Figure-20.

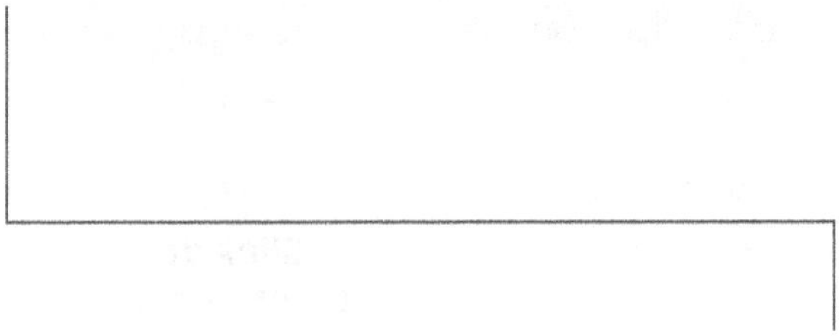

Figure-20. Wire to be drawn

Inserting transformer

- Click on the **Insert symbol** tool from the **Insertion** panel in the **Schematic** tab of the **Ribbon**. The **Symbol insertion** page will be displayed in the **Command panel**.
- Click on the **Other symbol** button from the page and select the single phase transformer as shown in Figure-21.

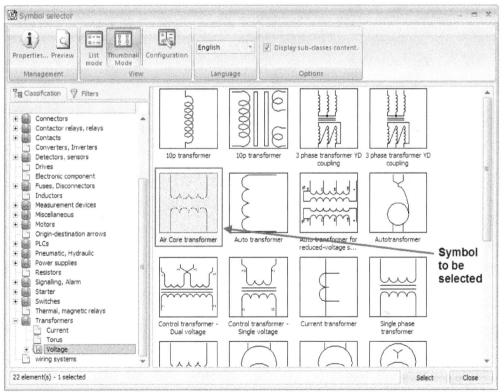

Figure-21. Symbol to be selected

- Click on the **Select** button from the dialog box. The symbol will get attached to the cursor.

- Rotate the symbol by 90 degree and place it by selecting the end point of the wire; refer to Figure-22.

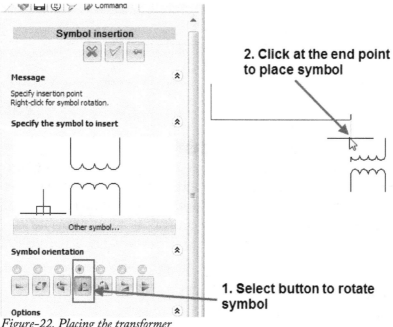

Figure-22. Placing the transformer

Drawing Rest of the Wires

- Again, click on the **Draw single wire** tool from the **Schematic** tab in **Ribbon**. The options related to wiring will display in the **Command panel**.
- Make sure that number of lines is set as **1** and **Wire style** as **230V AC** in the **Command panel**. Click at the open end point of the transformer and draw the wiring as shown in Figure-23.

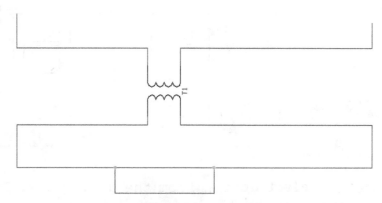

Figure-23. Wiring drawn

Inserting other symbols

*   Again, click on the **Insert symbol** tool from the **Schematic** tab in the **Ribbon** and one by one insert the symbols as shown in Figure-24.

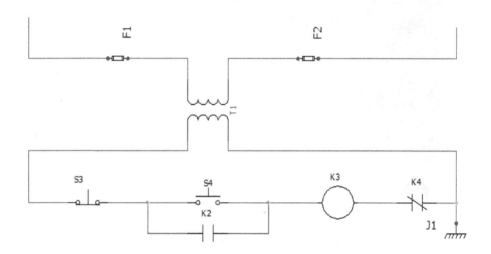

Figure-24. Inserting all components

Updating project data

Once we have done all the insertions, the last step is to update data of project which includes wire numbering, cable references, and so on.

*   Click on the **Update data** tool from the **Management** panel in the **Process** tab of the **Ribbon**. The **Wizard to update project data** dialog box will be displayed; refer to Figure-25.

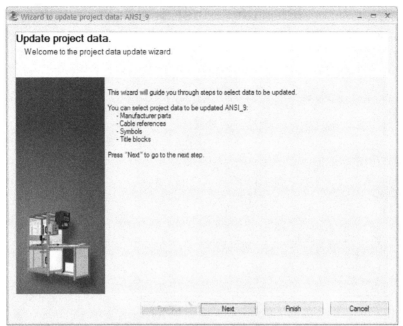

Figure-25. Wizard to update project data dialog box

- Click on the **Finish** button from the dialog box to update all the parameters.
- To display various markings like wire number; click on the down arrow next to **Show texts** button from the **Changes** panel in the **Schematic** tab of the **Ribbon**. Two tools will be displayed as shown in Figure-26.

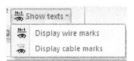

Figure-26. Tools to
display marks

- Click on the **Display wire marks** tool from the list. You are asked to select the wires for which you want to display the marks. Select all the wires by using the window selection and press ENTER. The wire numbering will be applied to the wires; refer to Figure-27.

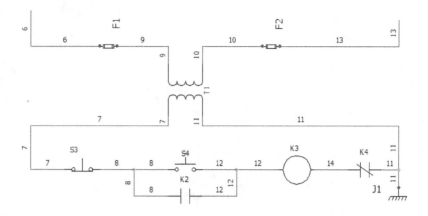

Figure-27. Schematic drawing with wire numbering

PRACTICE 5

On the basis of above practical, create the schematics of drawings shown in Figure-28 and Figure-29.

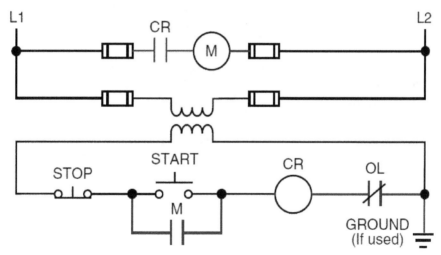

Figure-28. Practice 5

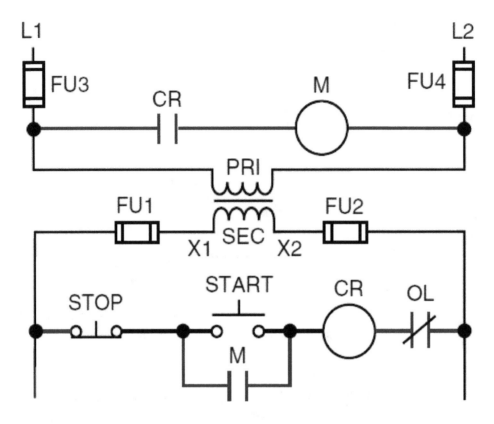

Figure-29. Practice 6

PRACTICE 6

Create panel drawing and bill of material for all the practical and practice questions discussed so far.

PRACTICAL 3

Create schematic for line diagram given in Figure-30.

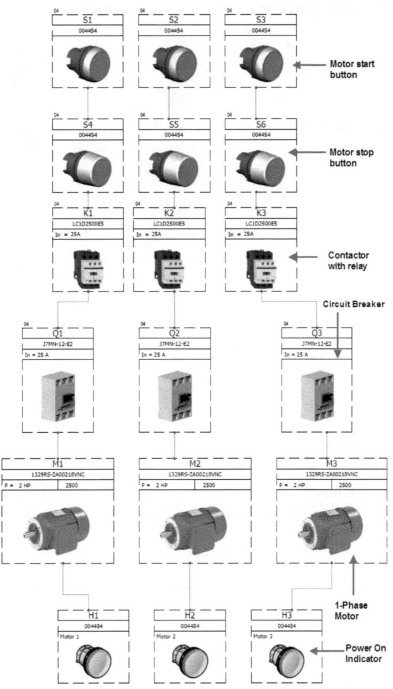

Figure-30. Practical 3

Starting a new Project

- Click on the **Projects manager** tool from the **SolidWorks Electrical** panel in the **File** tab of the **Ribbon** to display the **Projects Manager**; refer to Figure-31.

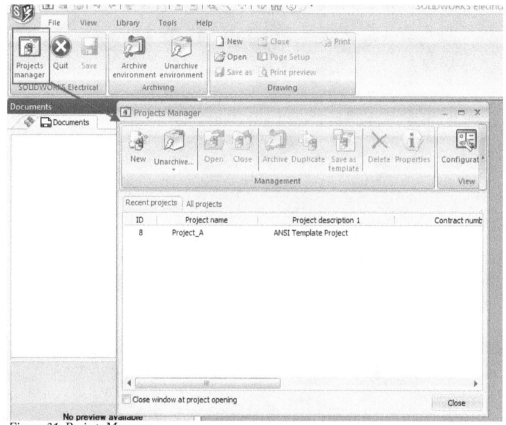

Figure-31. Projects Manager

- Click on the **New** button from the **Management** panel in the **Projects Manager**. You are asked to select the desired template from the **Create a new project** dialog box.
- Select the ANSI template and click on the **OK** button from the dialog box. The **Project language** dialog box will be displayed.
- Select **English** as language and click on the **OK** button from the dialog box. The **Project** dialog box will be displayed.
- Specify the user data as per the instructions given by your tutor.
- Click on the **OK** button after specifying the data, the newly created project gets added in the project list of **Projects Manager**.

- Make sure that the newly created project is open and then close the **Projects Manager** by selecting the **Close** button.

Creating Line Diagram

- Double-click on the Line diagram drawing from the **Documents Browser** in the left of the application window. The drawing will open.
- Right-click on the file name from the Documents Browser and select the **Replace** option in the **Title Block** cascading menu in the shortcut menu displayed; refer to Figure-32. The Title block selector dialog box will be displayed; refer to Figure-33.

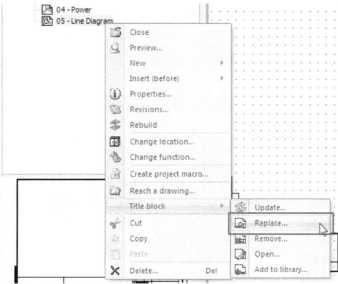

Figure-32. Replace option for title block

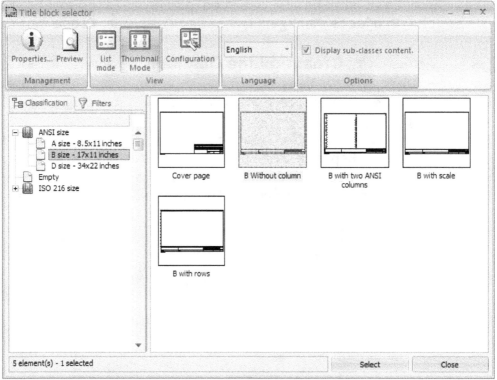

Figure-33. Title block selector dialog box

- We need a bigger size worksheet to create our line diagram, so click on the **D size - 34x22 inches** option from the left area of the dialog box. The title blocks with D size will be displayed on the right in the dialog box.
- Double-click on the **D without column without row option** in the dialog box. The template will change accordingly.

Adding Components in Line Diagram

- Click on the **Insert symbol** button from the **Insertion** panel in the **Line diagram** tab of **Ribbon**. The **Symbol insertion CommandManager** will be displayed; refer to Figure-34.

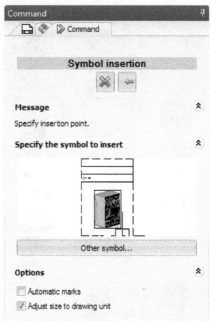

Figure-34. Symbol insertion Command-Manager

- By default, the symbol earlier used is displayed in the **CommandManager**. Click on the **Other symbol** button to display the **Symbol selector** dialog box.
- Click on the **Buttons, switches** category in the **Classification** tab at the left in the dialog box. The symbols related to buttons and switches will be displayed; refer to Figure-35.

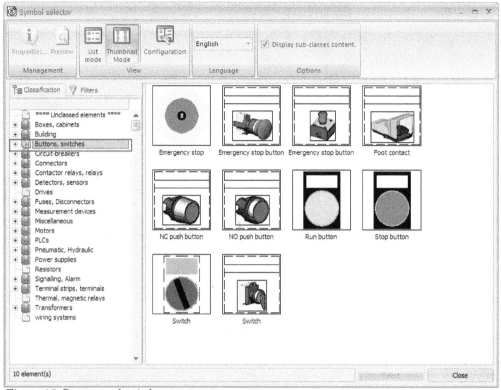

Figure-35. Buttons and switches

- Double-click on **NO push button** symbol from the dialog box. The symbol gets attached to cursor.
- Place the symbol near the top-left corner so that other symbols can be placed at the right and below this symbol.
- On placing the symbol, **Symbol Properties** dialog box is displayed; refer to Figure-36.

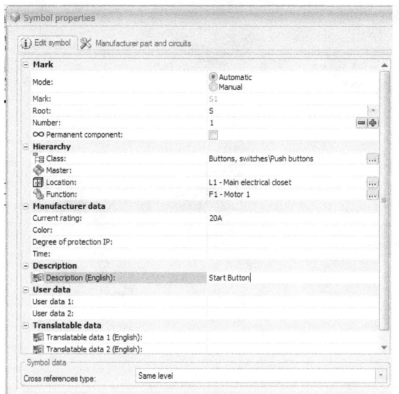

Figure-36. Symbol properties of button

- Set the current rating as **20A** and Description as **Start button.** Note that these values will come in the bill of material later. Now, we will specify the details of manufacturer of component so that it can be purchased in the market.
- Click on the **Manufacturer part and circuits** tab at the top in the dialog box. The dialog box will be displayed with options related to manufacturer.
- Click on the **Search** button at the left in the dialog box. The **Manufacturer part selection** dialog box will be displayed as shown in Figure-37.

Figure-37. Manufacturer part selection dialog box

- Set the number of circuits as **1** and number of terminals as **2** in the filters area; refer to Figure-38.

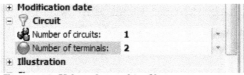

Figure-38. Values changed in filters

- Click on the **Search** button in the dialog box. Detail of manufacturer part will be displayed; refer to Figure-39.

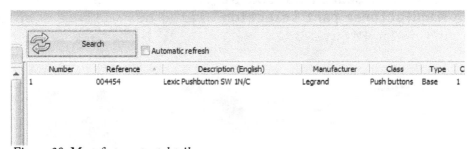

Figure-39. Manufacturer part detail

- Double-click on the manufacturer part from Legrand manufacturer and click on the **Select** button from the dialog box.
- Click on the **OK** button from the **Symbol properties** dialog box. The symbol will be added in drawing for manufacturer part.

Similarly, insert the other components for first motor circuit with descriptions as given next.

NC Push button: 20A (current rating), **Stop button** (Description), **Legrand** (Manufacturer), **004454** (Reference).

Contactor Relay: 25A (current rating), **Relay 1** (Description), **Schneider Electric** (Manufacturer), **LC1D25500E5** (Reference).

Circuit Breaker: 25A (current rating), **Circuit-breaker 1** (Description), **Omron** (Manufacturer), **J7MN-12-E2** (Reference).

Motor: 2HP (Power), **2500** (Speed), **Motor** (Description), **Allen-Bradley** (Manufacturer), **1329RS-ZA00218VNC** (Reference).

Signalling, Alarm/Luminous: 5A (Power), **Red** (Color), **220 V** (Voltage), **Motor ON** (Description), **Legrand** (Manufacturer), **004484** (Reference).

After putting all the parts, drawing should display as shown in Figure-40.

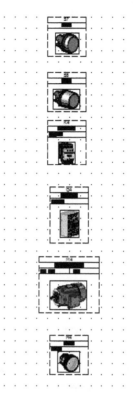

Figure-40. Symbols
placed in drawing

- Select all these symbols by using the cross-selection and press CTRL + C from the keyboard. All the symbols will be copied in system memory.
- Press CTRL + V from the keyboard and paste it two times as shown in Figure-41.

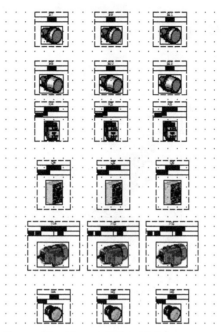

Figure-41. After pasting symbols

Connecting Cable

- Click on the **Draw cable** tool from the **Insertion** panel in the
Line diagram tab of **Ribbon**. The **Draw a cable CommandManager**
will be displayed; refer to Figure-42.

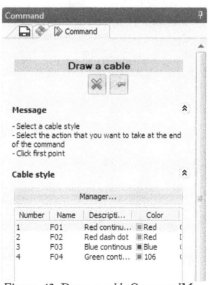

Figure-42. Draw a cable CommandManager

- Click at the bottom of S1 switch and then at the top of S4 switch; refer to Figure-43.

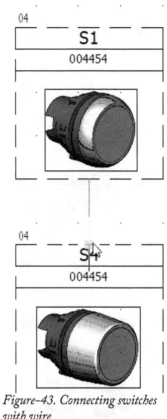

Figure-43. Connecting switches with wire

- Similarly, connect the other components in the line diagram; refer to Figure-44.

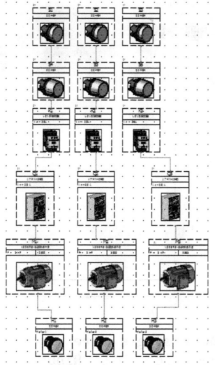

Figure-44. Connecting components in the line diagram

Based on this line diagram, we will now create schematic for the circuits.

Creating Schematic

- Double-click on the schematic drawing named **Power** from the **Documents Browser**. Blank drawing page will be displayed and tools related to schematic will be displayed in the **Ribbon**; refer to Figure-45.

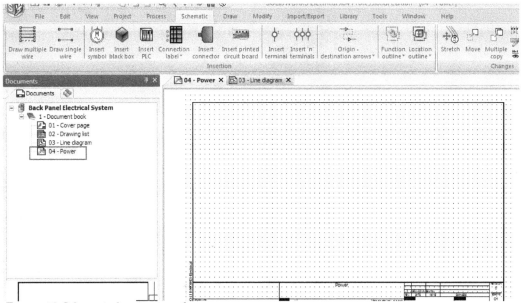

Figure-45. Schematic drawing interface

- Click on the **Draw single wire** tool from the **Insertion** panel in the **Schematic** tab of **Ribbon**. The **Electrical wires CommandManager** will be displayed; refer to Figure-46.

Figure-46. Electrical wires Command-Manager

- Click on the **Browse** button next to **Name** field in the **CommandManager**. The Wire style selector dialog box will be displayed; refer to Figure-47.

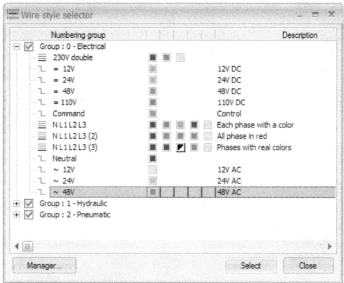

Figure-47. Wire style selector dialog box

- Click on the **Manager** button at the bottom of **Wire style selector** dialog box. The **Wire style manager** will be displayed; refer to Figure-48.

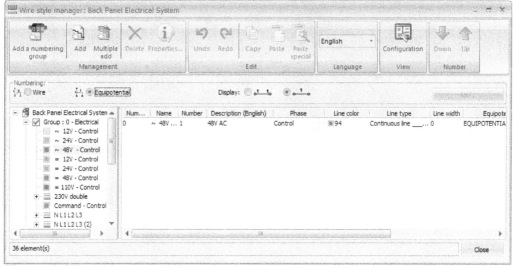

Figure-48. Wire style manager dialog box

- Click on the **Add** button from the **Management** panel in the dialog box. A new wire will be added in the list; refer to Figure-49.

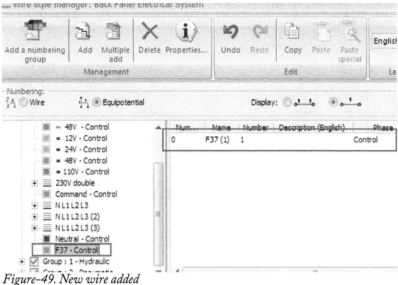

Figure-49. New wire added

- Right-click on the newly added wire and select the **Properties** option from the shortcut menu displayed; refer to Figure-50. The **Wire style** dialog box will be displayed; refer to Figure-51.

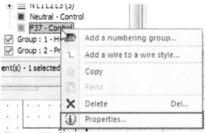

Figure-50. Properties option

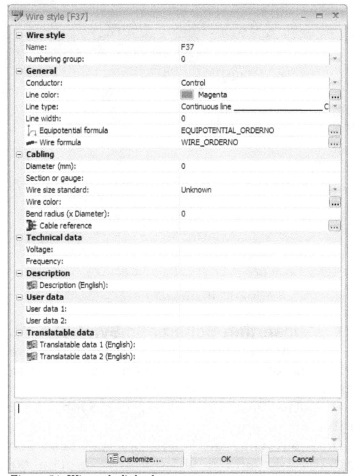

Figure-51. Wire style dialog box

- Click in the **Name** field and specify the name as **230V AC**. Similarly, set diameter as **2.5 mm**, Line color as **Red**, Wire size standard as **Section (mm²)**, Bend radius (x Diameter) as **25**, Voltage as **230 V** and Frequency as **60 Hz**.
- Click on the **OK** button from the dialog box. The wire will be created. Click on the **Close** button to exit the dialog box.
- Select the newly created wire from the **Wire style selector** dialog box and click on the **Select** button. The new wire will become active.
- Draw the wire as shown in Figure-52.

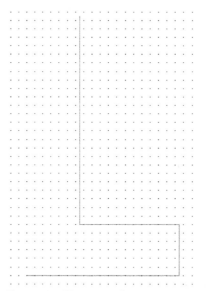

Figure-52. Wire created

Inserting Schematic Symbol

- Click on the **Insert symbol** tool from the **Insertion** panel in the **Schematic** tab of **Ribbon**. The **Symbol insertion CommandManager** will be displayed.

- Click on the **Other symbol** button, the **Symbol selector** dialog box will be displayed.

- Click on the **Buttons, switches** option from the left area of the dialog box and double-click on the **NO push button - Single Circuit - Momentary Contact** symbol; refer to Figure-53. The symbol gets attached to the cursor.

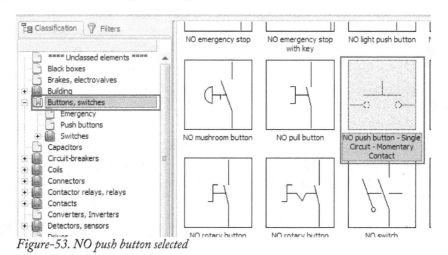

Figure-53. NO push button selected

- Click on the wire near the top end to place the symbol. The **Symbol properties** dialog box will be displayed; refer to Figure-54.

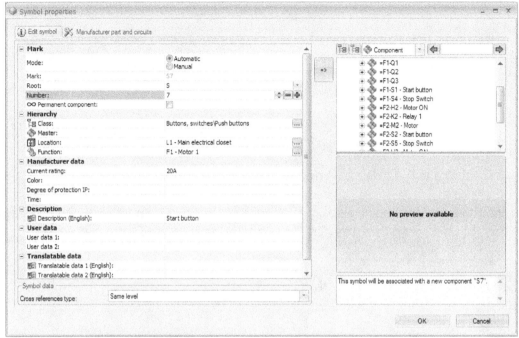

Figure-54. Symbol properties dialog box1

- Click on the **Same class** button above the **Component browser** in the dialog box; refer to Figure-55. The symbols of same class which have been used in the current project will be displayed.

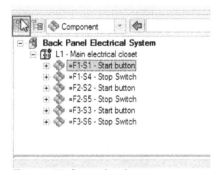

Figure-55. Same class button

- Click on the **S1-Start** button from the **Component Browser** and click on the **OK** button. The symbol will be added in the schematic drawing. Similarly, insert the NC push button matching its properties with S4 Stop switch; refer to Figure-56.

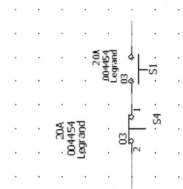

Figure-56. Stop-switch inserted

- Similarly, insert the other components for first motor circuit. Note that relay has two components in schematic relay contactor and relay coil. So, you need to insert both the components in the schematic; refer to Figure-57.

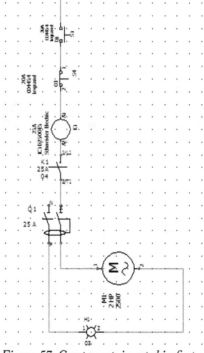

Figure-57. Components inserted in first motor circuit

- Draw a neutral wire connecting to the N terminal of circuit breaker in circuit; refer to Figure-58.

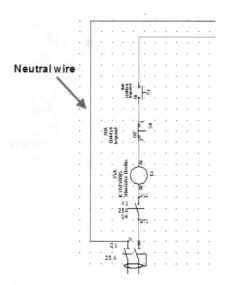

Figure-58. Neutral wire created

- Select this complete circuit by using window selection and make two copies of the circuit as shown in Figure-59.

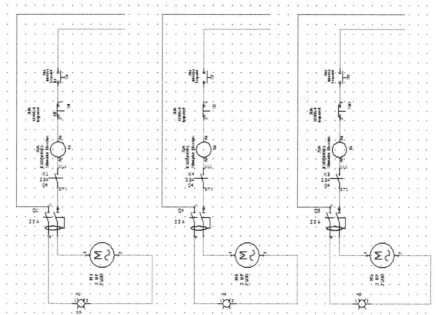

Figure-59. Circuit with multiple copies

Assigning Components to schematic symbols

- Right-click on the NO push button in second circuit and select the **Assign component** option from the shortcut menu; refer to Figure-60. The **Assign component CommandManager** will be displayed; refer to Figure-61.

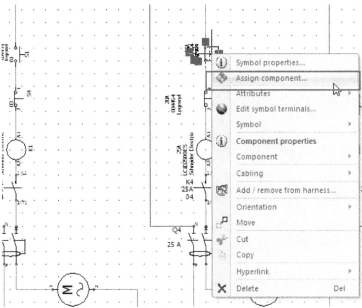

Figure-60. Assign component option

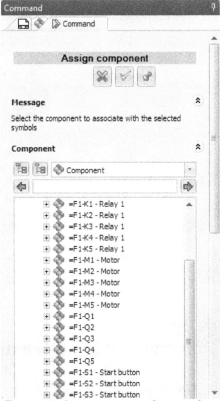

Figure-61. Assign component Command-Manager

- Select the **S2 - Start button** component from the **CommandManager** and click **OK** button. The component properties of S2 will be assign to the component and S7 will be removed from database.
- Similarly, assign the component properties to the symbols in circuit as per the line diagram; refer to Figure-62.

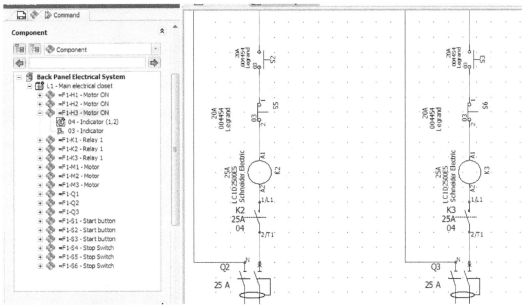

Figure-62. Assigning components to symbols as per line diagram

Connecting Terminal to wires

We are not given any power source in the question so we are going to connect all the open wires to terminals so that later power source can be directly connected to the terminals.

- Extend the wires using simple drag-drop functions on wires; refer to Figure-63.

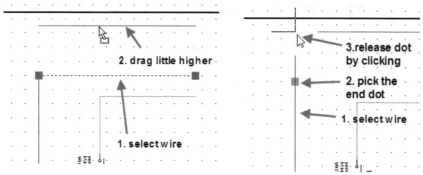
Figure-63. Drag-drop operations on wires

- After performing various drag-drop operations, make the wiring as shown in Figure-64.

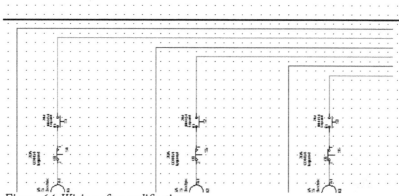
Figure-64. Wiring after modifications

- Click on the **Insert 'n' terminals** button from the **Insertion** panel in the **Schematic** tab of **Ribbon**. The **Terminal insertion CommandManager** will be displayed.
- Draw a vertical line intersecting the open ends of wires; refer to Figure-65. You are asked to define the orientation of terminals.

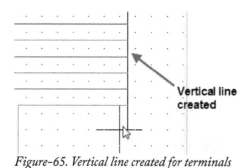
Figure-65. Vertical line created for terminals

- Click on the left of terminals. The **Terminal symbol properties** dialog box will be displayed; refer to Figure-66.

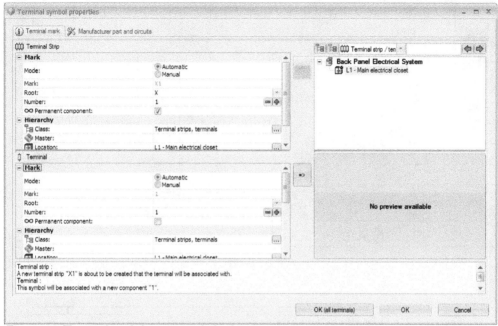

Figure-66. Terminal symbol properties dialog box

- Click on the **Manufacturer part and circuits** tab and select the **Entrelec** as manufacturer and Reference as **019532020** component from the list.
- Click on the **OK (all terminals)** button from the dialog box to create terminals. The schematic drawing will be displayed as shown in Figure-67.

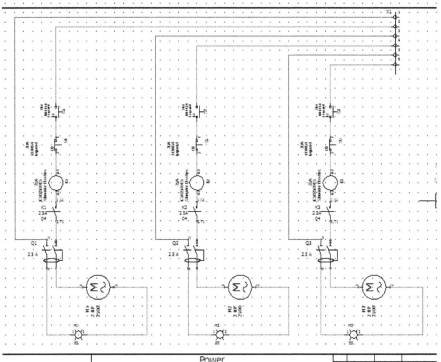

Figure-67. Drawing after adding terminals

PRACTICE 7

Create Line diagram, schematic, cabinet layout, SolidWorks Assembly, and bill of materials for motor circuit as shown in Figure-68, Figure-69, Figure-70, Figure-71, and Figure-72. Note: You will learn about creating SolidWorks Assembly in next chapter. Once you go through the chapter, you can come back and create the assembly file.

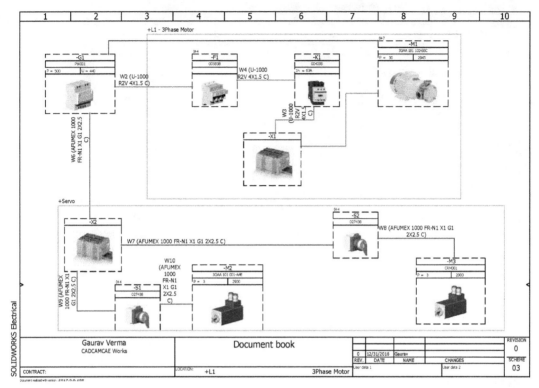

Figure-68. Line diagram

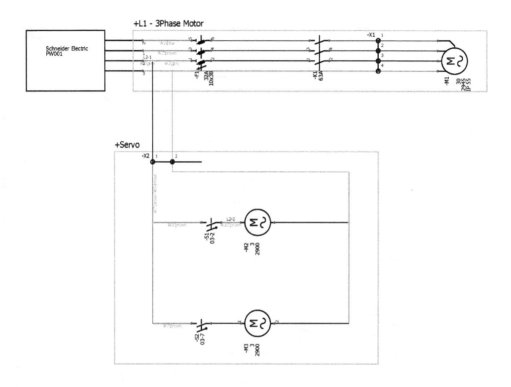

Figure-69. Schematic

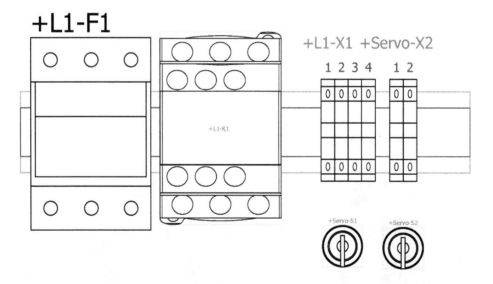

Figure-70. Cabinet layout

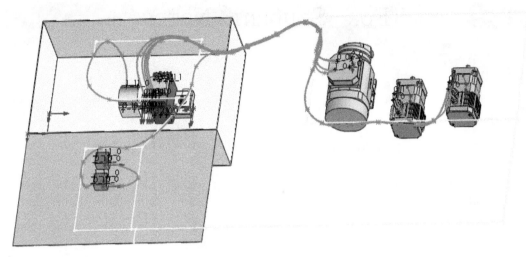

Figure-71. SolidWorks Assembly

ABB

Reference	Mark	Description	Quantity
3GAA 181 102-BBC	-M1	Motor 3-phases, type M3AA 180 LB ,2 poles - 400V VD - 50 Hz - High-output design, Flange-mounted	1

Crompton

Reference	Mark	Description	Quantity
CRM001	-M2 , -M3	Crompton 230V Single Phase Motor	2

Entrelec

Reference	Mark	Description	Quantity
010500220	-X1-1 , -X1-2 , -X1-3 , -X1-4 , -X2-1 , -X2-2	Simple terminal	6

Legrand

Reference	Mark	Description	Quantity
004095	-K1	Legrand Contactor 004095	1
005838	-F1	Mod TP 10x38mm Fuse Carrier	1
009213	-A2	Legrand Rail for Cabinet	1
035223	-A1	Stainless Wm Cab 600X400X2	1
027438	-S1 , -S2	Cam switch - changeover switch with off 45° - PR 17 - 4P - 20 A - screw fixing	2

Schneider Electric

Reference	Mark	Description	Quantity
PW001	-G1	Schneider Electric Power Supply	1

Figure-72. Bill of Materials

For Student Notes

FOR STUDENT NOTES

FOR STUDENT NOTES

Chapter 8
Electrical 3D

Topics Covered

The major topics covered in this chapter are:

- *SolidWorks Electrical 3D interface*
- *3D Electrical Parts*
- *Inserting Electrical Components*
- *CAD File Downloader*
- *Routing Wires*
- *Creating Routing Path*
- *Updating BOM Properties*

INTRODUCTION

SolidWorks Electrical Professional is combination of two interconnected electrical packages; SolidWorks 2D Electrical (Schematic) and SolidWorks 3D Electrical. Before this chapter, we have discussed about 2D part of SolidWorks Electrical. Now, we will discuss about 3D part of SolidWorks Electrical. Make sure that you have installed SolidWorks application and SolidWorks Electrical add-in for it. Before working with 3D electrical routing, we must have 3D parts that represent the symbols of SolidWorks 2D electrical.

CREATING SOLIDWORKS ELECTRICAL PART

Before creating any electrical part, you must have a model already created in SolidWorks. You can create a part or you can download it from 3dcontentcentral.com. Rest of steps are given next.

- Open the solid part that you want to make electrical part in SolidWorks; refer to Figure-1.

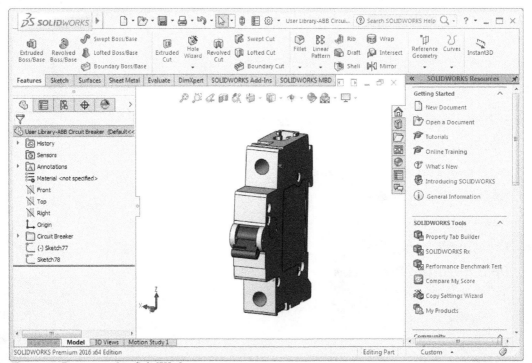

Figure-1. Part opened in SolidWorks

- Click on the **Add-Ins** option from the **Options** drop-down in the **Quick Access Toolbar**; refer to Figure-2. The **Add-Ins** selection box will be displayed; refer to Figure-3.

Figure-2. Add-Ins option

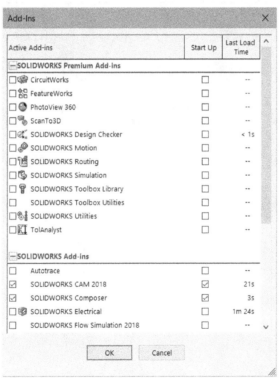

Figure-3. Add-Ins selection box

- Select the check box before **SOLIDWORKS Electrical** in the selection box and click on the **OK** button. The toolbar for SolidWorks Electrical 3D will be displayed; refer to Figure-4.

Figure-4. Solidworks electrical 3D toolbar

- Click on the **Electrical Component Wizard** tool from the **SolidWorks Electrical 3D** toolbar. The **Routing Library Manager** dialog box will be displayed with the **Routing Component Wizard** tab selected; refer to Figure-5.
- Select the **Electrical** radio button from the **Route type** area and click on the **Next** button. The **Select Component Type** page will be displayed; refer to Figure-6.

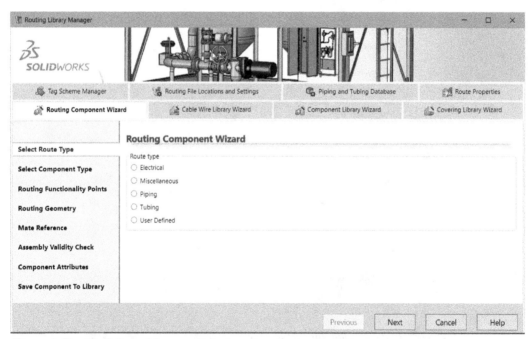

Figure-5. Routing Library Manager dialog box

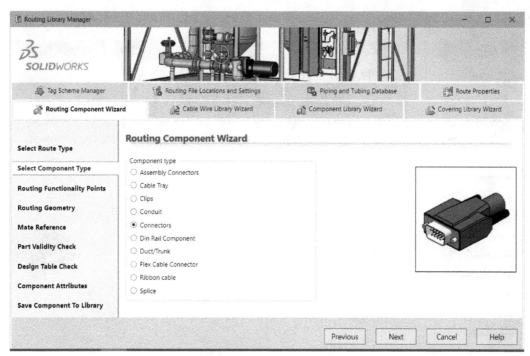

Figure-6. Routing Component Type page

- Select the desired option from the list. We are using the Connectors radio button because we want to create electrical connection points to a part. If you have a assembly component then you should use the Assembly Connectors radio button from this page. After selecting the radio button, click on the Next button. The **Routing Functionality Points** page will be displayed; refer to Figure-7.

- Select the **CPoint** radio button if you want to create connection point for wire without the related circuit of manufacturer information. Select the **CPoint with circuit information** radio button if you want to create wire connection points with related circuit information. This option is useful when your component is having many connection points for different circuits within it. Select the **CPoint from manufacturer part** radio button if you want to create connection points as per the manufacturer data. Select the **Cable CPoint** radio button if you want to create connection point for cables. We are selecting the **CPoint from manufacturer part** radio button in our case.

• After selecting the radio button, click on the **Add** button.

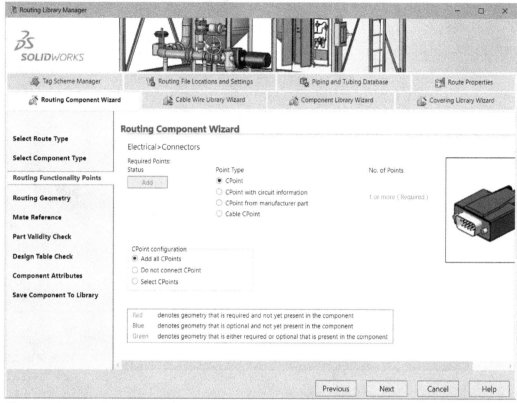

Figure-7. Routing Functionality Points page

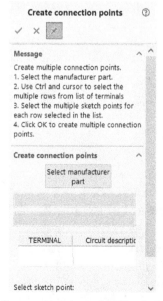

Figure-8. Create connection points PropertyManager

- Click on the **Select manufacturer part** button from the **PropertyManager**. The **Manufacturer part selection** dialog box will be displayed. Select the desired part using the options as discussed earlier in the book. The related terminals of the part will be displayed in the list in **PropertyManager**; refer to Figure-9.

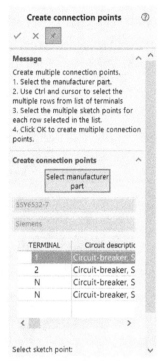

Figure-9. Terminals of part
displayed in PropertyManager

- Select the desired terminal from the list and click at the desired location on the face of model. The **SolidWorks Electrical** dialog box will be displayed; refer to Figure-10.

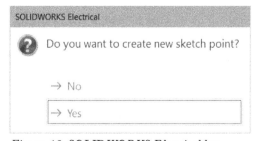

Figure-10. SOLIDWORKS Electrical box

- Click on the **Yes** button to create a sketch point for connection point and click on the **OK** button from PropertyManager. The connection point will be created and the **PropertyManager** will be displayed with one of the terminal boldface in the list.
- Similarly, create the other connection points; refer to Figure-11.
- Click on the **Keep Visible** (Pin) button in the **PropertyManager** and then click on the **OK** button from the **PropertyManager**. The connection points will be created and the **Routing Library Manager** dialog box will be displayed.

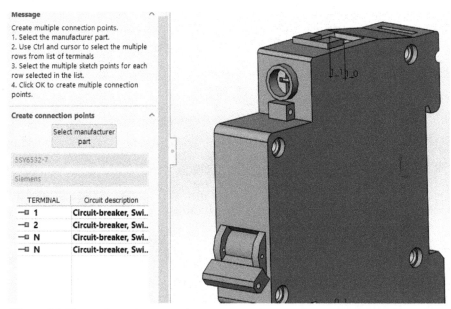

Figure-11. Connection points created

- Click on the **Next** button from the dialog box. The **Routing Geometry** page will be displayed. Set the desired parameters as required in the page and then click on the **Next** button. The **Mate Reference** page will be displayed; refer to Figure-12.
- Select the **For Rail** radio button if the component is to be connected with rail. Select the **For Cabinet** radio button if the component is to be connected to cabinet by its back face. Select the **For Cabinet Door** radio button if you want to place the component on cabinet door. In our case, we have selected the **For Rail** radio button.

- After selecting the radio button, click on the **Add** button. The **Create Mate Reference PropertyManager** will be displayed; refer to Figure-13.

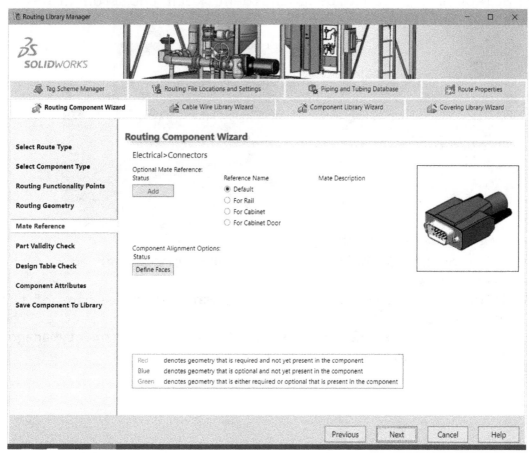

Figure-12. Mate Reference page

Figure-13. Create mate reference PropertyManager

- Select the face of component to be placed on the top face of the rail. You will be asked to select the front face of rail.
- Select the face to be coincident with the front face of rail; refer to Figure-14.

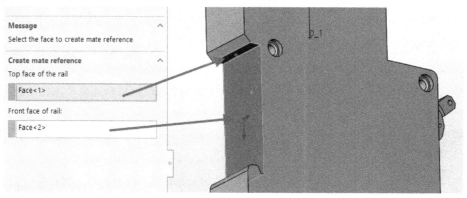

Figure-14. Faces selected for mate reference

- Click on the **OK** button from the **PropertyManager** to set the references.
- Click on the **Define** button from the **Mate Reference** page of the **Routing Library Manager** dialog box to set alignment of the part. The **Define all the faces of the component PropertyManager** will be displayed; refer to Figure-15.

Figure-15. Define all the faces of the component PropertyManager

- Select the left, right, top and bottom face of the component one by one and then click on the **OK** button from the **PropertyManager**.
- Click on the Next button from the dialog box to check if the part is valid or not. If there is any error shown then go back and rectify it otherwise, click on the Next button to modify component attributes.
- After setting the desired parameters, click on the Next button. The **Save Component To Library** page will be displayed; refer to Figure-16.

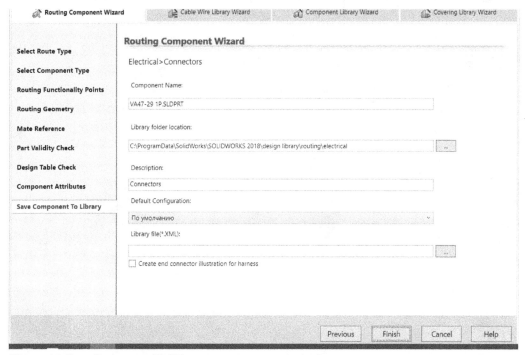

Figure-16. Save Component To Library page

- Set the desired location and component name to save the part. Click on the **Finish** button. **SOLIDWORKS** dialog box will be displayed asking you to save the part.
- Click on the Yes button from the dialog box. The part will be saved and a message box will be displayed.
- Click on the **OK** button from the message box.
- Click on the **Cancel** button from the **Routing Library Manager** dialog box and Click **Yes** from the dialog box displayed next to exit the dialog box. Click on the **Exit** button to close the dialog box.

CAD FILE DOWNLOADER

There is a big database of CAD parts from their manufacturers in electrical engineering available on internet. SolidWorks Electrical avails a tool named **CAD File Downloader** to use these manufacturer part files. The procedure to this tool is given next.

- Click on the **CAD File Downloader** tool from the **Tools > SolidWorks Electrical** menu in the Menubar; refer to Figure-17. The **CAD File Downloader** dialog box will be displayed; refer to Figure-18.

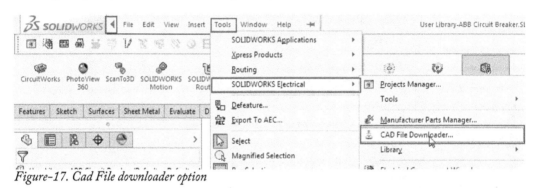

Figure-17. Cad File downloader option

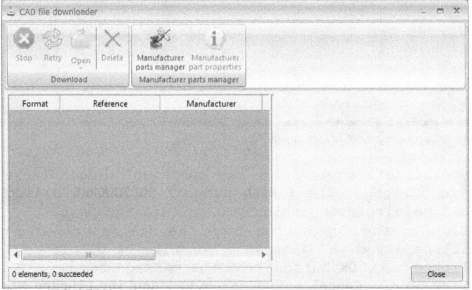

Figure-18. Cad File downloader dialog box

- Click on the **Manufacturer parts manager** button from the **Manufacturer parts manager** panel in the dialog box. The **Manufacturer parts manager** will be displayed; refer to Figure-19.

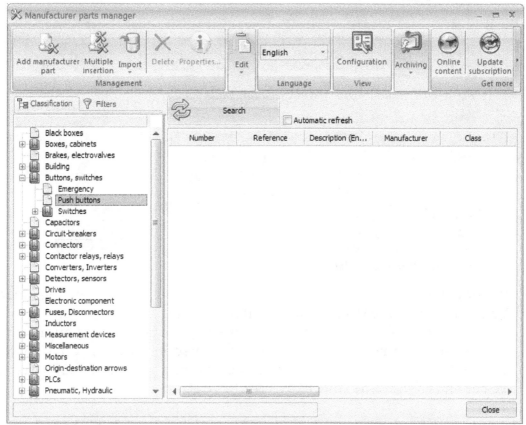

Figure-19. Manufacturer parts manager

- Select the desired category from the left in the dialog box and search the component for which you want to download the part file.
- Select the component from the table in the right and click on the **Download SOLIDWORKS 3D part** button to download the part; refer to Figure-20. The **CAD file downloader** dialog box will be displayed and SolidWorks will start looking for the part over internet. Once the part is found, it will start downloading the part file; refer to Figure-21.

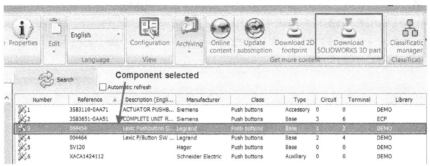

Figure-20. Download SOLIDWORKS 3D part button

Figure-21. Part downloaded

• Select the part from the dialog box and click on the **Open** button from **Download** panel. Now, you can check the part in SolidWorks; refer to Figure-22. Note that you might be asked to proceed with feature recognition, choose **No** in such cases because FeatureWorks can not identify mate references of SolidWorks Electrical.

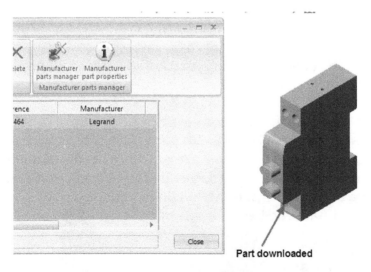

Figure-22. Downloaded part opened in SolidWorks

INSERTING COMPONENTS IN SOLIDWORKS ELECTRICAL 3D

In SolidWorks, you can insert the electrical components in the same way as you do in SolidWorks Assembly environment. But, there is also a special way designed to insert electrical components by using SolidWorks Electrical 2D projects. Before that we need to open the electrical project file in SolidWorks. The open project file is given next.

- In SolidWorks Electrical, open the desired project file and click on the **SolidWorks assembly** tool from the **Processes** panel of the **Process** tab in the **Ribbon**. The **Creation of assembly files** dialog box will be displayed; refer to Figure-23.

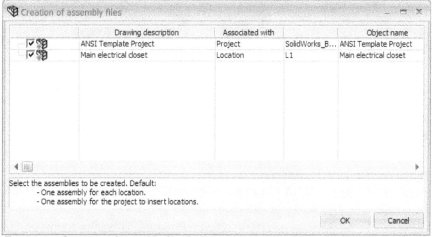

Figure-23. Creation of assembly files dialog box

- Select the check boxes for assembly files that you want to be add in project and click on the **OK** button. Blank assembly files will be added to the project.
- Start SolidWorks and add the **SolidWorks Electrical** Add-in by using **Add-Ins** dialog box as discussed earlier.
- Click on **Project Manager** tool from the **Tools** > **SolidWorks Electrical** menu; refer to Figure-24. The **Project Manager** window will be displayed; refer to Figure-25.

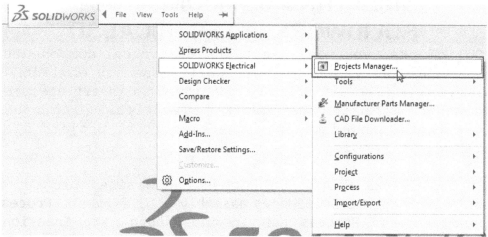

Figure-24. Project Manager tool

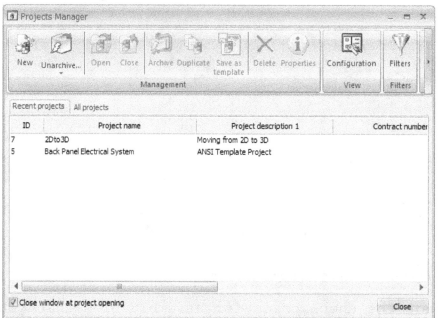

Figure-25. Project Manager dialog box

- Double-click on the project using which you want to create 3D panel drawing. The project file will open in **Electrical Project Documents** pane in the right of SolidWorks; refer to Figure-26. Make sure that you have added SolidWorks cabinet layout file in the project by using SolidWorks Electrical.

- Expand the nodes of project and double-click on the cabinet layout part file; refer to Figure-27. An assembly file will open and components of the project will be displayed in **Component explorer** at the right of SolidWorks; refer to Figure-28.

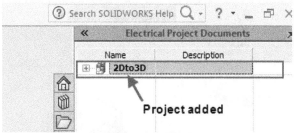

Figure-26. Project added in SolidWorks

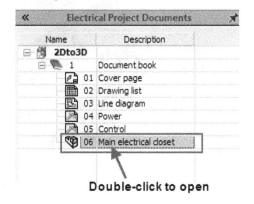

Figure-27. SolidWorks cabinet file

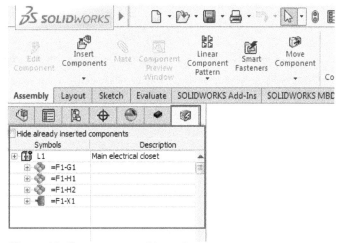

Figure-28. Components used in project

*   Expand the node for a component in the **Component explorer** and double-click on the part file. The component will get attached to cursor; refer to Figure-29. In some cases, you

may get a warning box like the one shown in Figure-30. In such cases, click on the **Insert the part** button from the box. If you do not want this warning box to be displayed then specify the size of component while defining its schematic/line diagram symbol.

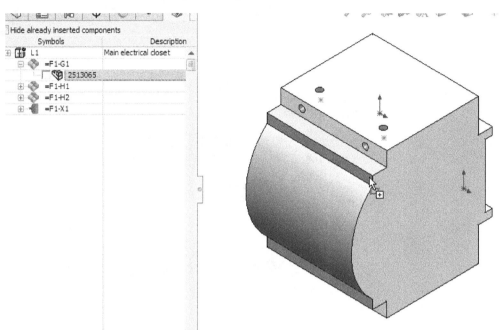

Figure-29. Component attached to cursor

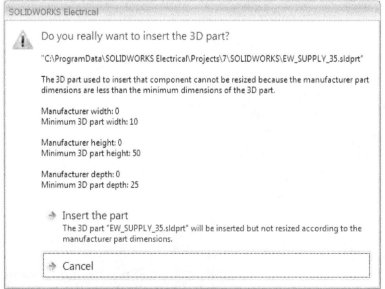

Figure-30. SolidWorks Electrical dialog box

ROUTING WIRES

SolidWorks Electrical has a tool named **Route Wires** to automate the process of wire routing based on the details provided in SolidWorks Electrical 2D. The procedure to use this tool is given next.

*   Click on the **Route Wires** tool from the **SOLIDWORKS Electrical 3D** tab in the **Ribbon**; refer to Figure-31. The **Route wire PropertyManager** will be displayed; refer to Figure-32.

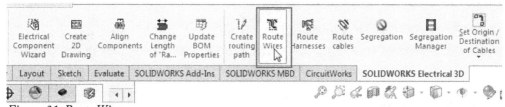

Figure-31. Route Wires

Figure-32. Route wires PropertyManager

- Select the **Show errors** check box from the **Routing Analysis** rollout in the **PropertyManager** if you want to check the errors generated while routing.
- Select the desired option from the **Select route** type rollout. If you want to create wires that look alike real then select the **SOLIDWORKS Route** radio button. If you want to generate splines of different color representing wires then select the **3DSketch Route** radio button. Note that selecting the **3DSketch Route** radio button greatly reduces the time required by system for performing automatic routing.
- Select the **Cable core follow routing path** check box if you want to make the cores of cable follow the routing path as their wires do.
- There are two options to use as renderer for routing, **Use splines** and **Use lines**. Select the desired option from the **Select renderer type** rollout. You can add tangency conditions at the links by selecting **Add tangency** check box from the rollout.
- Specify the routing parameters like deviation tolerance for distance between two routing paths or distance between connecting point and nearest routing path. These parameters are specified in the **Routing parameters** rollout.
- You can check the effect of specified parameters by selecting **Draw Graph** button; refer to Figure-33. Select the **Delete Graph** button to delete the graph.

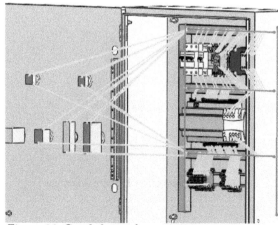

Figure-33. Graph drawn for routing

- From the desired algorithm from the **Shortest Path Algorithm** rollout and selected the desired engine from the drop-down displayed on selecting the algorithm.
- At last, click on the **OK** button from the **PropertyManager**. SolidWorks will start automatic routing and once the routing is complete, it will show errors/warnings of the routing via **Routing Analysis** window; refer to Figure-34.

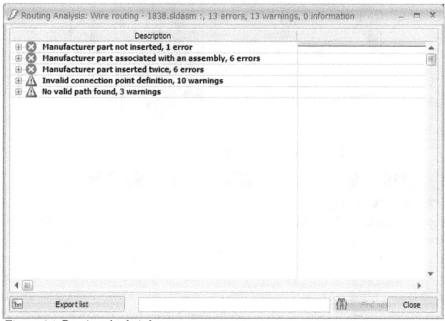

Figure-34. Routing Analysis box

- Expand each node and check the cause of error. Most of the time, these are simple errors like manufacturer part not inserted. In such cases, you need to insert the part. In some cases, you do not need a part to be inserted then you can ignore these errors. After performing modifications as per the errors, click again on the **Route wire** tool and perform routing.

Figure-35 shows an assembly after performing routing.

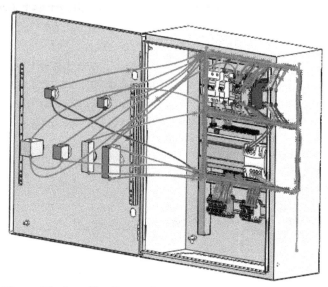

Figure-35. Assembly after routing

CREATE ROUTING PATH

The **Create routing path** tool is used to create path for routing wires. If we check the routing of assembly in Figure-35 then we can find that it is penetrating through the enclosure which is not generally considered in real-world; refer to Figure-36. To eradicate such situations, we use create routing path which tells system then wires should go through the specified path before making connection to the components. The procedure to use this tool is given next.

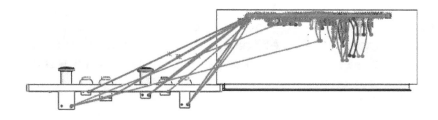

Figure-36. Top view of routing

- Click on the **Create routing path** tool from the **SOLIDWORKS Electrical 3D** tab in the **Ribbon**. The **Create Routing Path PropertyManager** will be displayed; refer to Figure-37.

Figure-37. Create Routing Path PropertyManager

- If you have already created 3D sketch for path in modeling area then select **Convert sketch** radio button from the **PropertyManager** and select the sketch. Click on the **OK** button from the **PropertyManager**. The routing path will be created.
- If you want to create a new routing path or you do not have an existing 3D sketch path then click on the **Create sketch** radio button and click on the **OK** button from the **PropertyManager**. A message box will be displayed telling you that only lines and sketch points can be used for creating routing path; refer to Figure-38.

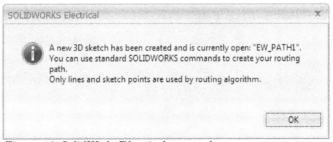

Figure-38. SolidWorks Electrical message box

- Click on the **OK** button from the message box and draw the 3D sketch; refer to Figure-39.

Lines drawn in 3D sketch

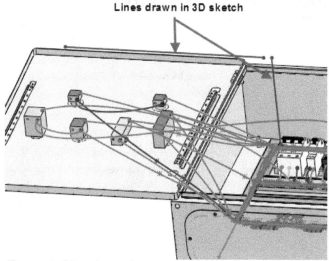

Figure-39. Lines drawn for routing path

- Exit the sketch environment. The routing path will be displayed in yellow color, by default.
- Now, click on the **Route Wires** tool from the **Ribbon** to check the difference caused in routing due to routing path; refer to Figure-40.

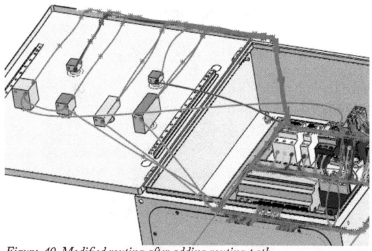

Figure-40. Modified routing after adding routing path

ROUTING CABLES AND HARNESSES

The **Route cables** and **Route harnesses** tools work in the same way as discussed for **Route wires** tool. Most of the options are same. There is a new option named **Update Origin/Destination** check box in case of routing cables which enables to update the origin and destination data of cable in the 2D schematics based on 3D routing and vice-versa.

UPDATE BOM PROPERTIES

The **Update BOM Properties** tool is used to update the bill of material by adding the cable/wire length and other parameters in reports based on the properties in 3D electrical model. The procedure to use this tool is given next.

- Click on the **Update BOM Properties** tool from the **Ribbon**. The Bill of Material will be updated automatically and a message box will be displayed; refer to Figure-41.

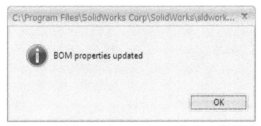

Figure-41. Message box stating updation of BOM

- Click on the **OK** button to exit.

In the same way, you can use the **Align Components** tool to align components and **Change Length of "Rail" or "Duct"** tool to change the length of rails and ducts in the assembly.

The tools like Create 2D Drawing from assembly are native tools of SolidWorks. You can find the details of SolidWorks modeling and assembly tools in our other book, SolidWorks 2018 Black Book.

PROJECT

Create schematic diagram and 3D assembly of motor control circuit in which there are two push buttons to start and stop the motor. Glow bulbs should be there in the panel to denote whether the motor is running or stopped. Three glow bulbs will be there for showing whether electrical supply is active or not. Note that circuit should be safe from electric surge.

Hint for Schematic is available as pdf in book's resource kit. Write at cadcamcaeworks@gmail.com to get the resource kit.

Index

Other Books by CADCAMCAE Works :

SolidWorks 2018 Black Book
SolidWorks 2017 Black Book
SolidWorks 2016 Black Book
SolidWorks 2015 Black Book
SolidWorks 2014 Black Book

SolidWorks Simulation 2018 Black Book
SolidWorks Simulation 2017 Black Book
SolidWorks Simulation 2016 Black Book
SolidWorks Simulation 2015 Black Book

SolidWorks Electrical 2017 Black Book
SolidWorks Electrical 2016 Black Book
SolidWorks Electrical 2015 Black Book

AutoCAD Electrical 2018 Black Book
AutoCAD Electrical 2017 Black Book
AutoCAD Electrical 2016 Black Book
AutoCAD Electrical 2015 Black Book

Creo Parametric 4.0 Black Book
Creo Parametric 3.0 Black Book

Creo Manufacturing 4.0 Black Book

MasterCAM 2017 for SolidWorks Black Book
MasterCAM X7 for SolidWorks 2014 Black Book

Autodesk Inventor 2018 Black Book

Autodesk Fusion 360 Black Book

www.ingramcontent.com/pod-product-compliance
Lightning Source LLC
Chambersburg PA
CBHW080550060326
40689CB00021B/4809